A
PHYSICIST'S PERSPECTIVE
ON
GOD

A

PHYSICIST'S PERSPECTIVE

ON

GOD

ROADMAPS TO WISDOM
THROUGH SCIENCE AND LIFE

ALAN TAI, PhD

XULON PRESS

Xulon Press
2301 Lucien Way #415
Maitland, FL 32751
407.339.4217
www.xulonpress.com

Printed in the United States of America.

ISBN-13: 978-1-6305-0875-3

TABLE OF CONTENTS

Endorsement . . . . **vii**
Preface . . . . **ix**
Part 1: Introduction 1
   Ch. 1: My Background to Realize the Wisdom in Science
      and Life 3
   Ch. 2: My Path to Pursue the Wisdom in Science and Life 7
   Ch. 3: Communication Through the Scientists and their Lives 14
Part 2: The Wisdom Communicated Through Invisible Qualities. .17
   Ch. 4: Unseen Things 19
   Ch. 5: Law of Physics. 22
   Ch. 6: Universal Order 25
Part 3: The Wisdom Communicated Through Eternal Power 39
   Ch. 7: In the Beginning. 41
   Ch. 8: The Vastness of the Universe 48
   Ch. 9: The Uniqueness of the Earth and Humans. 53
Part 4: The Wisdom Communicated Through Divine Nature 65
   Ch. 10: Love and Light. 67
   Ch. 11: Spirit and Wisdom 78
Part 5: The Wisdom Communicated Through Special Revelation .87
   Ch. 12: The Bible 89
   Ch. 13: Jesus Christ. 101
Part 6: Conclusion 115
   Ch. 14: The Combined Roadmap .116
   Ch. 15: Triune God and You 123
Bibliography 129
Appendix I 131
   General Communication Via www.scienceandlife.net 131
Appendix II 134
   Introduction of T Transform for Mathematical Fraction 134
Acknowledgments 137
About the Author. 139

ENDORSEMENT

I am writing a glowing endorsement of Dr. Alan Tai's book, "A Physicist's perspective on God" Dr. Tai studied under my tutelage in another discipline. He is a careful and conscientious thinker. He has good academic intuitions and his conclusions are the result of sound logic and methodology. Dr. Tai's thesis is one I fully embrace. Science and faith are not strange bedfellows. The findings of science are not to be feared by people of faith, but considered as evidence of God's creative hand. This book goes beyond the standard apologetic and presents a personal journey that is mindful of the best findings of science. The integration of science and faith is a new frontier and one that is long overdue. Rationalism, naturalism and secularism have tipped the scales for far too long. We need science-based scholars with a strong faith conscience to right the scientific scales and guide our thinking to a worldview that is consistent with reality as we know it while being faithful to the tradition of biblical revelation. Scientific fundamentalism is a real threat to academic integrity. It impoverishes a thinker and leads to false conclusions. Dr. Tai's book creates new possibilities for a scientific reader. It is worthwhile for any serious student of the bible and science. Dr. Tai's speaks from experience, not from a position of religious fundamentalism. Read it, all of it, you will be helped.

-Rev. Dr. Robert Hunter,
Adjunct Professor at Grand Canyon University,
Creator & Owner at RIP'd 4 Life.

Christian faith and science seem contradictory. This is only a misunderstanding if we do not have a deep understanding of the two. The Bible is full of scientific records, such as: the creation of God in Genesis, the description of nature and stars in Job, etc. Of course, the most important revelation for the Bible is to record the attributes, will, and salvation of God.
I am grateful to have Alan Tai, Ph.D. in writing his book "A Physicist's Perspective on God: Roadmap to Wisdom Through Science and Life". This book shares his in-depth scientific knowledge and understanding of faith,

and investigation of the relationship and how they complement the two. This is not just a book of "apologetics". It gives us a deeper understanding of how science and nature are in harmony and related to the creation processes revealed in the Bible. This valuable reference book can give us a deeper understanding of our faith and help us to share more supportive evidences and scientific facts with non-believers.

Rev. Stephen W. Leung,
Founding Pastor of Remembrance of Grace Church,
(+5 churches planting) in Hong Kong.

Alan's story and perspectives have long inspired me and encouraged me in my own faith. I've been privileged to know and work with Alan for over 30 years. Through his outstanding creativity and scientific understanding, he achieved breakthroughs in diagnostic medical imaging, which were described in multiple patents. Knowing his story, I was excited to read his book, and look forward to the impact it will have on others who have a love for science and questions about God. He brings a uniquely objective, unbiased, and scientific perspective through his personal story and scientific observations. He came to the U.S. for undergraduate school as an atheist with a passion for math and physics. But before he became an accomplished physicist, he became a passionate believer, who saw the Creator and Christ evidenced in beautiful ways by science.

Steve Miller, Senior Vice President,
Engineering, Seno Medical.

PREFACE

Dear fellow human being,

No matter where and when you live, study, and work, people around you should respect one another regardless of their backgrounds, races, beliefs, and ages. If we can understand more about other people's situations and promote dialogue among different cultures, the world will have more peace and fairness. I like to read books about lives like those of Marie Curie, Albert Einstein, Isaac Newton, and people with varied backgrounds and beliefs. You can explore this interest of mine on my website www.scienceandlife.net.

Growing up in Hong Kong, where people describe the city as "East meets West," I experienced and learned many different cultures and beliefs. I have always loved learning about science and scientists' lives. The lives of scientists motivated me to pursue the Truth. While reading their biographies about their discoveries in science, I have enjoyed the dialogue provided in their journeys through science and life.

I am a scientist, not a politician, so I try to stay away from political topics. The politicians usually offer complicated and complex points of view that can be biased and subjective. For anyone who thinks they have a good point about a certain political argument, there exists another thinker with an opposite argument describing a completely different outcome. In reality, though the intent of a political stance may be good for some people, it might feel evil to others. History and current world news already demonstrate peace is not easy to achieve without mutual respect and understanding.

Science itself is more geared to objective thinking and studies that deal with data collection and analysis. "It is a branch of knowledge or study dealing with a body of facts or truths systematically arranged and showing the operation of general laws. In general, science is related to a systematic knowledge of the physical or material world, gained through observation and experimentation."[1] Scientists usually keep an open and objective mind about whether what they are studying is considered truthful, and realize

---

[1]   https://www.dictionary.com/browse/science, accessed Oct 18, 2019

it must be subjected to research to be verified as true. They can hypothesize, assume, and even imagine certain theories to be tested and checked by experiments to determine if they are correct and accurate. The goal of the contribution to science is to claim nothing false or containing mistakes within the scope of the scientists' investigation. Once their theories prove wrong, partially or completely, they should make every effort to correct and acknowledge their mistakes and give respect to others' insights. This embodies the spirit of studying science. Not only striving to give credit to others' contributions, the scientists know they always stand on their predecessors' shoulders when accomplishing anything significant.

With this spirit of studying science in mind, I want to share the processes of finding the wisdom in different aspects of science and life. My understanding about how God, the Creator, communicates to human beings, opens my spiritual eye to appreciate His divine wisdom and realize His beauty.

This book reviews why the following comparisons are true based on the divine wisdom from the Creator. By looking at the design of the heavens and earth, I understand God's creation and wisdom is superior than any human's creativity and wisdom.

Divine wisdom (DW) >> (exceeds) human wisdom (HW)

And, divine wisdom (DW) >> (exceeds) artificial intelligence (AI)

The Bible also shows a similar argument, per the following comparison: "For the foolishness of God is wiser than men, and the weakness of God is stronger than men."[2] Even as AI has recently arisen with new technologies to expand the capabilities of HW, it still has limits when compared with DW.

The general definition of wisdom is related to the "quality of having experience, knowledge, and good judgment; the quality of being wise."[3] People are considered to have wisdom when they make good use of their experience, knowledge, and good judgement to impact others and themselves wisely. The Bible further describes genuine wisdom: "Yet among the mature we do impart wisdom, although it is not a wisdom of this age or of the rulers of this age, who are doomed to pass away. But we impart a secret and hidden wisdom of God, which God decreed before the ages for our glory."[4]

---

[2] I Cor. 1:25, NIV Bible, unless otherwise indicated.

[3] https://www.lexico.com/en/definition/wisdom, accessed Dec. 3, 2019

[4] I Cor. 2:6–7

Can we find this hidden wisdom, so we can enhance our wisdom to be a wiser person? We encounter absolute truth and relative truth in our daily living. The absolute truth, like the laws of physics, can make us wise when we know how to apply them in science and technology to improve human living. The relative truth, like theories and hypothesis, are subjected to verification and tests to see if they are correct or not. Any immature use or assumption of the relative truth as always correct can lead to a negative impact on or even a disaster for human life. An example: certain races are considered to be superior to other races because they evolved better during the natural selection processes. Hitler made use of these assumptions in raising war against other countries and eradicating any race, including Jews, that didn't match his "ideal." His actions resulted in one of the worst disasters in human history. That's why we need to be careful and spend time studying if "truths" are correct or not.

Let's use the example of a judge: all the evidence needs to be presented before making a decision in the court. The judge also must know and understand the related law for the case before he or she can apply the law objectively to announce the verdict. Discussing the topic of God is a sensitive matter to some people. If you can temporarily put aside your predetermined ideas or concepts toward religions, you assume the position of a judge and can review the presentation of evidences with respect to some universal laws.

In this case, I as a physicist and a scientist, want to provide you with the following presentations related to the communication of wisdom in science and life. Regarding universal laws, it is assumed to be within the scope of the scientific investigation. After reviewing the presentation objectively, you can make your own decision whether to accept or reject the final verdict of the case. I respect you as a human being, whether you are a believer or nonbeliever about what I present here. At the end of the day, you are responsible for the choices and actions that lead to the final outcome of your life.

I saw on television a true testimony from a person who shared how he spent time investigating in-depth the religion of his parents. He did not want to simply follow in the footstep of his parents without knowing the ultimate Truth. After seeking and checking the beliefs of his parents, he came to firmly believe his choice of religion made sense. He considered it Truth instead of superstition. The time he spent in the spirit of pursuing the Truth meant a lot to him, so he did not regret his decision for his own life and his beliefs. This kind of person needs to be respected because his belief is backed up with careful study and review about what the Truth is. My hope is you will make a wise decision like this after reviewing the amazing wisdom behind science and life.

My mission is to write a book that comprises my life journey in pursing the Truth, which is the key to opening the door to the mysteries in life. I also dedicate this book to all those who love the Truth and are willing to pursue its path.

There are numerous resources and books regarding science, faith and wisdom that you can access on the Internet, and in libraries and bookstores. The unique portion of this book is related to the "roadmap" that can give you some insight and direction about connecting with the divine wisdom. The general definition of "roadmap" is a blueprint in achieving a specific goal through different phases of processes and implementation. It is an effective presentation using graphic and visual display to connect the key tactics and thoughts so as to communicate high-level strategical steps to the readers.[5] Viewing a picture of a completed jigsaw puzzle, you will find that the picture facilitates you to put the scattered pieces of the puzzle together into a completed jigsaw puzzle. Similarly, the function of a roadmap is like a picture to help you to re-organize the puzzle of the Biblical truth and to apprehend the divine wisdom behind the writing of the Scripture. With this basic tool of "roadmap" on hand, you may able to break through an obstacle in communication with God. Your willingness in seeking the biggest treasure in life, of knowing God, is still needed. I hope this book will provide a valuable map to point you in the right direction, setting you on the path of a treasure hunt to find the ultimate Truth in life. You will find the Bible is capable of lighting up this path.

I was a former technical leader in a scientific role with General Electric, Healthcare division. Every year, the leaders met to go through the latest roadmaps of new products slated for the coming years. The roadmaps were usually simple and easy to understand, so all the leaders had a common tool and similar insight into a medical device.

Now, my new mission in life is "impacting life in time that last for eternity" I hope the roadmaps in this book can contribute by sharing communication wisdom to the readers. Once you open your heart to listen to the messages sent out by God, a path can be established to connect the communication channel and find the wisdom.

If you are not sure about God's existence, there is no better way than pursuing your own realization of how God communicates to you. Consider the

---

[5] https://www.dictionary.com/browse/roadmap, accessed Dec 2, 2019

analogy of a radio receiving a clear communication channel even through all the airwaves' background noise. This proven scientific concept provides strong evidence of the existence of this kind of divine channel, which can reach you directly. You will determine God is real in your life when you receive His communication to you. I pray your path of finding and knowing God will enrich your life abundantly because you have a heart of seeking the Truth.

If you are a person who likes to see the bottom line first, you can skip some of the chapters. I recommend you go to chapter six for the presentation of "Universal Order." It presents a scientific map for finding the most valuable treasure in life. You can use the map and direction from chapter thirteen to see the wisdom communicated through the "Special Revelation" in life.

May your journey of finding the ultimate Truth be filled with peace and joy. May your seeking heart open the door to see the beauty and wisdom behind the creation, that is, the Creator. All blessings and wisdom flow from God, who created the entire universe so we can see the beauty in science and life.

I wish you the best in seeking divine wisdom.
Respectfully,
Alan Tai
October 22, 2019

PART ONE

INTRODUCTION

Chapter One

MY BACKGROUND TO REALIZE THE WISDOM IN SCIENCE AND LIFE

I grew up in a traditional Chinese family in Hong Kong. My ancestors on my father's side, the Tai family, were among the ethical group named "Hakka (客家)" who had resided in Hong Kong for at least a few hundred years. They used to worship the heaven and the heavenly queen by offering incense, chicken, and roast pork, etc., to idols. When I was young, before the age of ten, my family lived next to a temple called "Heavenly Queen" temple — "Tin Hau" temple in the Chinese pronunciation — near Causeway Bay in Hong Kong. According to the tradition passed through the generations of my father's ancestors, an altar was mysteriously found in the coastal area near Causeway Bay, where the "Tin Hau" temple was located, more than one hundred years ago. People thought this altar was sent to them by the "Heavenly Queen" so they could offer their worship to her. By giving certain sacrifices, people received protection and blessings from Heaven.

At that time, those who lived near Causeway Bay were farmers, and fishermen who went to the ocean for fishing. Life could be dangerous when the fishermen encountered life-threatening storms and strong winds. The farmers could also lose their harvests because of strong winds damaging their crops.

Every year, quite a few storms passed through Hong Kong, and some of them could make a huge impact in damaging the environment. Technologies were not available yet to warn of the storms. Fisherman relied mostly on experience to look up to the skies to predict if their journeys out to sea would be safe or not. They had no confidence in their interpretations of the sky alone, and wanted to reply on spiritual or heavenly persons to guide their lives. Heavenly Queen seemed to be merciful and powerful enough to help and protect them when people worshipped her and other heavenly persons.

My ancestors on my father side, were involved in building the "Tin Hau" temple, and managed the worship ritual of the Heaven Queen and other spiritual hosts.

The big event at that time was the birthday of the "Heavenly Queen." The people celebrated her birthday by bringing an offering of money to the temple staff. Artists and Kung Fu disciples performed music, drama, and a costumed lion dance through the week of celebration. From time to time, I went inside the temple to check it out. Various idols sat in different locations of the temple. Behind the main altar, located close to the back of the temple, two large idols stood about ten feet high. The idols wore red dresses and featured black faces. They looked somewhat scary to a young boy like me at that time. I did not have a good impression of the environment, especially with more idols in corners of the temple. I specifically remember an idol of the monkey king, a fictional character in a famous novel *Journey to the West,* on the left corner of the temple.

All of the idols and statues inside the temple seemed to communicate to those who followed the traditions and believed the stories of protection and blessings that would pass on to them. In addition to the environment of "fear," other processes expressed the promise of "help" to the worshippers. People prayed to the statues for blessings and asked for direction. Many wanted to learn about their futures and obtain some solutions related to their problems in life. So, they knelt in front of the idols and whispered their questions for usually less than ten minutes, while holding a lot of bamboo sticks etched with separate numbers.

Once a stick was picked from the lot, they brought it to an "interpreter" who explained the meaning of the ticket corresponding to the number of the sticks. Tickets were printed with poems about certain historical stories. Based on what "seekers" desired, the interpreter used the story in the ticket to compose some guidelines related to the seeker's life issues. In some cases, the interpretations seemed to predict their futures. Of course, the "seekers" would likely share with their relatives and friends the correct predictions of the fortune telling through the spirits because of their sincere worship of the idols. My mother was among the seekers of the fortune telling from Heavenly Queen.

My ancestors on my mother side, the Lun family, lived in Mainland China, not too far away from Hong Kong. The location sat in the southern part of China in Canton province within 150kms northwest of Hong Kong. My mother, Lee Siu-King, born during World War II, was adopted by another family at a young age because of difficult circumstance in the Lun family. She was raised separately from her family and never got a chance to go to school. She always had a heart for learning and taught herself reading and writing Chinese even without a formal education. It was a happy ending for the Lun brothers and sisters when my uncle, Lun Lam Kwan, found my mother during a visit to the "Tin Hau Temple." The reunion of my mother to the Lun family was a turning point for my parent, my three brothers, and myself. All of a sudden, I had more uncles, aunts, and cousins. In addition, I found out more about the legend of my mother's ancestors.

The history of the Lun family can be traced back to the Ming dynasty of more than 500 years ago. A few scholars in the Lun family achieved the special honor of ranking among the top three in the highly competitive examinations offered by the Imperial officials in China. One famous incident related to a poem written by one of the Lun Scholars, Lun Wenxue (倫文敍, year AD 1467 to 1513), about a famous painting: "The Hundred Birds Returning to the Nest."[6] The poem appeared to be simple, but turned out to be profound when interpreted with some mathematical calculation. The first half of the poem was, "Heavenly born of one bird after one bird, three four five six seven eight birds."「天生一隻又一隻，三四五六七八隻。」

The interpretation is interesting: $1 + 1 + 3 \times 4 + 5 \times 6 + 7 \times 8 = 100$

So, the above equation turns out to be "one hundred," matching the theme of the painting. The second half of the poem related with a phoenix and birds given special meanings that compare the "righteous" official and "common" officials in the Imperial ruling of the people at that time: "So few phoenix but so many birds, pecking on ten million stones of the people."「鳳凰何少鳥何多？啄盡人間千萬石。」

The poem was like a parable that communicated the wisdom and experience of the writer. But the readers at that time were not able to apprehend the wisdom of the poem right away. They even got very disappointed by its common appearance and seemingly nonsense numbers until the writer explained about the real meaning behind the numbers. They then realized the wisdom communicated through the numbers and language in the first half of the poem. The poem further noted that, in Chinese pronunciation, the nest, or "巢", was close to "朝," which meant the dynastic for imperial ruling.

---

[6] Painted by Su Shi (Chinese: 蘇軾, AD 8 January 1037–24 August 1101, general name: Dongpo, Chinese: 東坡), was a Chinese calligrapher, painter, poet, politician, and writer of the Song dynasty. (蘇東坡真跡名畫「百鳥歸巢圖」)

Lun Wenxue used the birds in the second half of the poem to represent the "common official" in the dynastic at that time. They were squeezing the people to get money by means of corruption and greediness. This parable expressed the need for more phoenixes, corresponding to the righteous and talented officials. In general, the poem gives the readers a picture of the real situation behind people's lives; a roadmap toward more trustworthy intellectuals in the government. This inspirational story also motivates people, while interpreting the real meaning behind a wise person's writing; to pay more attention to the wisdom and revelation behind the writing.

Although I have a traditional Chinese family, my individual personality is to pursue wisdom in mathematics and science. I was fascinated by the beauty of mathematics and its application which drew my attention to pursue the discovery of the true meaning of life.

Chapter Two

MY PATH TO PURSUE THE WISDOM IN SCIENCE AND LIFE

My mother's brothers were also successful businessmen, who opened consumer product factories in Hong Kong and in the Canton province. During the summers of my high school years, my older brother, Ming, and I worked in two of my uncles' factories: Lun Fat Kwan and Lun Lam Kwan. That experience also helped me understand what the grassroots factory workers' life was about. I appreciated the diligences of the factories' owners and workers in working together to contribute to the successful development of the economy of Hong Kong.

During the 1960s and 1970s, as I grew up in Hong Kong, the economy there steadily increased as the influential center of Asia, and its political situation was relatively stable. Across the border of a small river north of Hong Kong, the "closed door" policy at that time created certain barriers for communication from western development in science and technology to mainland China. In addition, the "culture revolution" at that period made people's life more difficult in learning, working, and contributing to the society.

But, on our side, the environment was completely different. Hong Kong people enjoyed a high degree of freedom for learning, working, and contributing to the success of Hong Kong as one of "the four Asian tigers or dragons" in Asia. Hong Kong was also nicknamed the "Pearl of the Orient" that shone among the countries in the world. During the "Golden Years" of Hong Kong development, I was motivated to learn science through frequent visits to libraries to borrow and read books related to science and life.

The Internet was not there yet, so the main sources for learning science came through books and studying in schools. I loved books and still feel good when I visit bookstores and libraries. I can somehow "communicate" with the scientists through their biographies and stories related to their path of development in science and life. Actually, my "dialogue" with scientists was based on me diving into the description of their life stories and processes. This involved their childhood years, when their scientific minds were molded; their midlives, when their discipline, maturity, and imagination generated breakthrough theories and experiments; and their later lives when they contributed to the community without boundaries and passed on the torch of scientific innovation to the next generation.

On the other hand, my attempt to communicate with "God" or a "spiritual being" started with the environment I was born into. There were different kind of religions with different types of temples and church buildings in Hong Kong. The high volume of people visiting certain temples of Eastern religions were usually due to "word of mouth" and traditions of the people who encountered certain blessings in the past or expected certain blessings in the future. Attending a church and believing in Jesus are usually considered to be westernized. Traditional Hong Kong people, include my parents, think of Jesus as a foreigner from Europe or a western civilization.

My experiences with religions in Hong Kong also included the three different religion-associated schools in which I studied. I entered the first grade in an elementary school associated with Buddhism. At that time, Hong Kong was a British colony and enjoyed a high degree of freedom in religion. Different religious organizations and charities sent funding and resources to start elementary and high schools in Hong Kong so the students could learn the related religions in their studies. My impression of Buddhism was related to statues of the Buddha, which appeared to be smiling and deep thinking. The history of Buddhism can be traced back to certain philosophies about life that did not start with idol worshipping. The statues of Buddha were added as different branches of Buddhism evolved in different times and places. Failing in most subjects except mathematics, I only studied in that school for first grade.

Next, I entered a Catholic school, St. Antonio Catholic elementary school, where I studied from second grade to sixth grade. The school was located directly above the house I lived in, in Causeway Bay, the locale of the "Heavenly Queen" temple. In this school, I first heard about the stories of the Bible. Statues of St. Mary, Jesus, and the apostles were placed in different locations around the school. They appeared to be more friendly than those I saw inside the "Heavenly Queen" temple. People passing the worship hall got a sense of solemnness in addition to the atmosphere of certain unspeakable "love" throughout the campus of the school. I also enjoyed the Christmas party when all the students could watch movies and received appealing gifts and candies.

I recall a young and lovely female teacher who converted to be a Catholic nun. Her enthusiasm in teaching and faith in her religion expressed a certain genuineness of caring for people behind her dedication. The feeling of a holy environment and prayer during the mandatory Mass—the Catholic worship service—also drew my attention and changed my attitude toward "western religion." Even though my family tradition of the "Eastern God" did not endorse any involvement of the "Western God," I started to pray to St. Mary by memorizing her prayer as I did in the school.

I remember one story taught by the nun related to the special power of St. Mary. During the final judgement, all the work during the lifetime of a person was put on a scale to be determined if he or she was good enough to enter heaven. The handkerchief of St. Mary, who heard the prayer of the person, would be put on one side of the scale. This would change his or her status from bad to good, because the tears of St. Mary somehow made the difference. I did not understand why, or what really happened after people died. Drawn by the religion's charm and virtues, I liked to participate in the Mass, and engaged in some religious activities in the Catholic Elementary School. However, I did not have a relationship with God through Jesus Christ. My so-called "faith" at that time was in a shady ground that would not stand when trial and difficulty came.

My results in academic examinations were not so good in elementary school, so in one school year, I needed to see the headmistress, a Catholic nun. I still remember she looked straight into my eyes and asked me what caused my grades to be poor. I answered that I watched too much television and did not spend much time studying. She then asked me to promise her not to watch television till my academic performance improved. I had no choice but to say "yes" to her. I did try not to watch TV after the meeting with her. I studied behind the TV because the living space was small in Hong Kong. I could hear the TV sound and the temptation was too strong for me to withstand. I struggled somewhat, but gave in and went around to the front of the TV and enjoyed the TV programs. My academic performance in Chinese language for entering middle school did not meet the minimum requirement of the Hong Kong Education department. Therefore, I did not receive an assignment to a middle school, per my academic performance.

I needed to find a middle school on my own since I did not get the subsidized tuition from the government. That was certainly a trial to me, especially because I prayed to St. Mary quite often. I was wondering, *Why didn't St. Mary or God answer my prayer?*

Had I been admitted to another Catholic middle school and passed the middle school entrance examination, I might have become a Catholic person. When I look back at this kind of question, I ask myself if my faith was based on favorable things happening at a specific time or on the Truth, if it pointed to a longer period of time?

Because I was not effective enough in my language memorization to deal with the language examinations, I would miss the chance to further my education unless a private school accepted me. I needed to approach different middle schools and take their tests to determine if I still qualified. A new middle school and high school up in the hill from where I lived was an "Islamic College." I was so desperate to be accepted in a school, I did not really care what kind of school it was or what religion was associated with the school. I managed to pass the Islamic College entrance test and started my student journey in a "new religion" school.

I had no idea what kind of religion related to the Islamic College, but I was glad my education could continue. I discovered the related religion did not belong to Westerner, nor Easterner. It belonged to Middle Easterners, as far as I could understand. All the students were required to attend a religious class about the Koran once a week. The religious education leader, Mr. Pao, was a Muslim scholar educated in the Middle East who could pray in the Arabic language during all student assemblies. His prayer included some songs with melodies I heard, but I did not understand the meaning of the prayers. From my observation, Mr. Pao sincerely prayed to communicate some messages to God (Allah), the essential points of which he also tried to share in our religion class. He was a patient and dedicated man, and the only Muslim educator who taught Koran classes to all students of all grades.

Most of what I learned in the Koran class was related to Muslim law, teaching people not to do wrong and evil things, and telling people about Allah. I felt sorry when I brought pork to school for lunch as eating pork was not allowed. One time, I forgot to study for the final examination. Students failed the Koran class would not be promoted to the next grade. Nervous during the examination time, I tried hard to write down good things about Allah and just mentioned some "common sense" answers to different religious and moral questions. I passed the Koran examination and was able to move up one grade in Islamic College.

During my middle and high school years, I got confused about religions: the Koran included stories of Abraham, who was also taught in my Catholic

Bible class in elementary school. It made me wonder if all religions had a common source. The religions I encountered always guided people to do good. If people did not follow these religious teachings about doing good, they would be punished for doing evil things. How could I know which religion was the right choice? I did not see an immediate need for any of them.

Life after high school was busy as I personally endeavored to focus on more studies and working. I had failed the English language subject in the Hong Kong standard test and could not continue to study for university attendance. So, I started working a series of low-level jobs. However, I was eager to learn science and technology, and studied at night at a technical institute. Science seemed to be the only truth worth pursuing.

Maybe evolution theory could answer where humans came from. Even though I wished there was a God who created all things, I pondered how I would know and communicate with God. Nevertheless, it was not in my heart to seek religions to find God. My impression of religions at the time was mainly based on my observation of people who regularly attended church or temple. True believers needed to give enough money and time to be considered religious people.

I would ask, "Why would I invest my time and money in some uncertain endeavor?" Especially when my dream of seeking wisdom in science and life was my highest priority. If I were a good enough person, I might not need religion. Evolution theory showed how people evolved from animals to human beings. I was an atheist at that time. I believed nothing but science to be the final Truth of life.

My family and relatives were supportive of my science education, because Confucian philosophy and Chinese tradition valued education favorably among other virtues. The same uncle who found my mother, Lun Lam Kwan, knew I was eager to continue my education. He and other relatives encouraged me to apply to universities abroad. Knowing they trusted and cared for my learning spirit, I deeply appreciated their love, not to mention their financial support of my education. If you review the biographies of Isaac Newton and Albert Einstein, you will find their uncles were among key family members who motivated their pursuits toward scientific education when they were young. The amazing achievements of these scientists—my role models—were mostly developed when they were young and introduced to new ideas and scientific wonders.

I worked and studied for some time after high school. Then, in August 1981, I had the opportunity to come to the United States to study. My horizons were opened and I was prepared to discover more about science and life. I decided my major purpose in pursuing higher education was to know science. I made up my mind not to spend time in any religious activity. I wanted to focus my time in studying physics and to discover its universal laws, which also fascinated Einstein.

I finally fulfilled my promise to the Catholic headmistress in elementary school to give up watching TV to spend more time studying. I basically did not watch TV for seven years while studying in the United States. I was able to make the choice because studying physics was more interesting than watching TV during my undergraduate and graduate student years.

Pursuing my dream of becoming one, I read a lot of biographies about the lives of scientists. Their passion and interest in science motivated and encouraged me to follow in their footsteps regarding physics. In particular, Albert Einstein, Isaac Newton, and Marie Curie drew my attention so vividly that I considered them as my science teachers and friends. Sometimes I wished I could communicate to them, even though their time on Earth had long gone. Their persistence and wisdom in understanding the scientific truth opened the door to classical and modern physics that pushed the science and technologists forward by at least a few decades. They are my role models as scientists who improved the lives of humankind.

I love science and physics. The more the scientific knowledge I gain about the law of physics, the more I can gain wisdom in using and applying them in the advance of science and technologists. There are so many mysteries in this universe. I wish I could get a glimpse into understanding the original cause and source of this world through science. How did this universe come into existence and become what it is now? What is the Big Bang Theory all about? Does the Big Bang Theory give us more evidence about creation or evolution as many scientists think? Is there a God that created the wonderful heavens and Earth? If there is a God for the first cause of these mysteries, how can I communicate with God so that I may know Him?

Galileo called nature the "Book of Nature." Mathematics is the language of this book. Just like wanting to read an English storybook, you may want to know the language in order to know the meaning behind the book. Therefore, the wisdom of nature is best understood through mathematics or its related principles. The current mathematical formation of nature and science results from the contributions of many wise mathematicians and scientists.

The theories of mathematics can be verified and applied to facilitate the establishment and advancement of physics. Mathematics itself is not bounded by space and time. The same mathematical truth can always be used and checked in different locations and periods. It is a universal language and can be studied by people without boundaries. Human language can change or evolve over time in different societies, but mathematical truths stay the same in the past, present, and future.

If there is a divine being that created the world as it is now, does He use mathematics as a foundation to engineer the wisdom of creation? Or, in human terms, is God also a mathematician besides a Creator? The maker of this world could follow some natural law we may know, and also some supernatural law human beings do not know yet, to form and shape the universe and all material within.

I visited my cousin in Toronto, Canada, at the end of 1981. He shared with me the Christian gospel and brought me to various church meetings. I also saw my cousin changed into a more caring and sincere person. He used to be a naughty and goofy person before he became a Christian. That week with him became the turning point of my journey to a new life. My heart opened and became willing to read God's Word. I met Jesus Christ and know that He is real. I accepted Him as my Savior and Lord. I found Jesus touching my life more than any other religion. Through Jesus Christ, I found a way to build a relationship with God, the Creator. It was a life-changing experience.

One night in 1982, while studying the Bible, the Spirit of Truth touched me gently and amazingly. Jesus Christ inspired me with the Truth I'd sought throughout my whole life. This divine message came to me in an instant, like a vision my spiritual eye saw and a word my ear heard. My instinct and response were to kneel to worship the great God. The triune God revealed to me that we can discover the blueprint of His creation and salvation plan. I even wrote down my revelation in a Bible and vowed to share this treasure in the proper time, by His Grace. This special connection to my Lord Jesus was so vivid I still remember it clearly even after more than thirty-six years.

Consequently, I found the hope in Christ worthy of time spent to know His Word and experience His guidance. Communicating with God through Jesus Christ, I gained the genuine wisdom to break the barrier that separated God and men. It also transformed my life.

Chapter Three

COMMUNICATION THROUGH THE SCIENTISTS AND THEIR LIVES

I n AD 1623, Galileo published *The Assayer* dealing with the scientific studies of comets. He made some of his most famous and remarkable methodological declarations when he came up with the "Book of Nature."[7] This religious and philosophical concept looks at nature as a figurative book to be studied for knowledge and understanding, and "written in the language of mathematics."[8] Galileo believed divine wisdom was embedded within this book of nature and God is known by nature in his works, and by doctrine in his revealed word.

The scientific investigations for Galileo's concept have been ongoing since the beginning of human civilization. To understand the wisdom of this concept, we need patience, an objective mind, and the pursuit of truth. Otherwise, if people used their own subjective minds to explain it, the outcome could lead to misunderstanding and even misuse. For example, taking the observation of gravity in nature. The falling motion of two objects are compared: if one is heavy and one light, a few hundred years ago the general public assumed the heavier weight would move faster than the lighter object. However, Galileo was determined to analyze this problem scientifically. He did experiments to compare the falling time for the two objects at the same distance. He found the falling time for both the heavy and light objects to be the same.

After Galileo, Isaac Newton showed and explained this phenomenon with the law of universal gravitation. This scientific discovery and other ongoing investigations in physics opened a new period of classic mechanics in the eighteenth century. When people came to understand and apply this

---

[7] https://plato.stanford.edu/entries/galileo, accessed May, 2019

[8] ibid

wisdom in science and technology, a new chapter of civilization began with the "Industrial Revolution."

In AD 1905, Albert Einstein, selected by *Time* magazine as the "most influential person" in the twentieth century, published a paper of "special relativity" at the age of twenty-five. This breakthrough discovery in physics enabled human beings to further understand nature from a new perspective of time versus space, and energy versus matter. This new chapter of the book of nature, related to modern physics, caused people to revise their traditional thinking about science.

All three scientists had one thing in common: they all believed God is the Creator and author behind this book of nature. As stated by Newton in *The Principia*, aka *Mathematical Principles of Natural Philosophy*, "This most beautiful system of the sun, planets, and comets, could only proceed from the counsel and dominion of an intelligent and powerful Being."

Throughout Newton's life, he passionately pursued an understanding of the book of nature and also the "Book of Books" (Bible). Newton believed the Bible communicated the divine revelation from God to humans using language they could comprehend. The time he spent studying the Bible was compatible with the time he spent studying science, as illustrated by his writing of 1.3 million words on biblical subjects.[9] Knowing the Creator was the author of the Book of Nature and the Book of Books, Newton was inspired by divine wisdom during his studies of science and life.

Newton's life is best summed up in the epitaph on his tomb in Westminster, London. "Here is buried Isaac Newton, Knight, who by a strength of mind almost divine, and mathematical principles peculiarly his own, explored the course and figures of the planets, the paths of the comets, the ideas of the sea, the dissimilarities in rays of light, and, what no other scholar has previously imagined, the properties of the colors thus produced. Diligent, sagacious and faithful, in his expositions of nature, antiquity and the Holy Scriptures, he vindicated by his philosophy the majesty of God mighty and good and expressed the simplicity of the gospel in his manners..."[10]

Einstein further explained his studies by humbly stating, "We are in the position of a little child entering a huge library filled with books... The child dimly suspects a mysterious order in the arrangement of the books,

---

[9] https://www.christianitytoday.com/history/issues/issue-30/faith-behind-famous-isaac-newton.html

[10] https://themathematicaltourist.wordpress.com/2013/09/14/isaac-newtons-tomb/

but doesn't know what it is. That, it seems to me, is the attitude of even the most intelligent human being toward God. We see a universe marvelously arranged and obeying certain laws, but we only dimly understand these laws. Our limited minds cannot grasp the mysterious force that moves the constellations."[11]

Numerous scientists, who laid the foundation of modern science, believed in God. "For most medieval scholars, who believed God created the universe according to geometric and harmonic principles, so science — particularly geometry and astronomy — were linked directly to the divine. To seek these principles, therefore, would be to seek God."[12] Between the years AD 1901 and 2000, a majority of the Nobel laureates had association with religions and believed in God.[13] Their human wisdom in science witnessed the existence of a supernatural being that created the universe with divine wisdom.

Finding the right channel to communicate with God is a very important mission that cannot be ignored. Let us continue our investigation from the perspective of "general" revelation in science and "special" revelation in the life of Jesus Christ.

---

[11] http://www.alberteinsteinsite.com/quotes/einsteinquotes.html

[12] https://en.wikipedia.org/wiki/European_science_in_the_Middle_Ages, accessed June, 2019

[13] Baruch Shalev, 100 Years of Nobel Prizes, Americas Group, 2005: 59; 65.4% or 427 people were Christian, winners of Nobel Prizes including sciences, economics and peace. 21.1% or 138 people were Jewish. 3.2% or 17 people were other religions including Buddhists, Muslims, Hindu. 10.5% or 68 people were Atheists, Agnostics and freethinkers.

PART TWO

THE WISDOM COMMUNICATED
THROUGH INVISIBLE QUALITIES

Chapter Four

UNSEEN THINGS

J ust as there are a few assumptions in establishing the theory of quantum mechanics, we take some assumptions in reviewing how the existing universe came to be as it is now. Subsequent experiments test and check if the theory can be validated. These steps embody a typical approach in the scientific world. We apply similar approaches and arguments in understanding the origin of this universe. One assumption is: "God created the universe and the world." We also assume God's word, the Bible, reveals and tells us something about God and His works.

Romans 1:20 starts this specific approach, as this Bible verse gives a general description from both the creation point of view and the scientific point of view. "For since the creation of the world God's *invisible qualities*—his *eternal power* and *divine nature*—have been clearly seen, being understood from what has been made, so that people are without excuse[14]. Whether you agree with God's revelation or not at this moment, we take this assumption first and expand from these three specific descriptions related to the creation of the universe: 1. God's invisible qualities, 2. His eternal power, and 3. His divine nature.

Later in this book, I will provide perspective on the evidence and technical data to back up this assumption.

First, what are invisible qualities?

Isaac Newton wrote, "When I look at the solar system. I see the Earth at the right distance from the Sun to receive the proper amounts of heat and light. This did not happen by chance."[15] In this example, the visible parts are the Earth, Sun, light, etc. The invisible parts or qualities are the heat—infrared light is unseen by the human eye—as well as the proper amounts of heat and light. For the Earth to receive these, it somehow needed to be

---

[14] Romans 1:20

[15] https://www.brainyquote.com/authors/isaac-newton-quotes, accessed Oct 2019

placed the right distance from the Sun. Otherwise, the temperature on Earth would not be suitable for complex life to survive, and the light reaching Earth would not be proper for plants to grow by photosynthesis.

The questions are: "Who put the Earth at this proper distance?" "Did it happen by God or by chance?" The more we understand the solar system, the better we see the picture of needing a very precise location away from the Sun for the life on Earth to survive. Here, science provides a very helpful tool of mathematics and physics to screen out the false or incorrect theories and assumptions.

Meanwhile, science can provide scientific evidence to rule out the existing Earth's location could be possible due to chance or random expansion. The latest scientific theory about the origin of the universe is the "Big Bang theory," which will be discussed more in Chapter 7, "In the Beginning." The solar system was the result of an expansion, while the Earth's distance away from the Sun was one of the sequential expansion events. Mathematicians and physicists can calculate the related probabilities for an object with a mass like the Earth and a source with a mass like the Sun. Assumptions of certain physical models need to be made for this calculation or simulation. Other factors, like Earth's atmosphere to retain air—including oxygen, nitrogen, carbon dioxide and so on—can also affect the temperature of the Earth's surface for human survival. In any case, we need a starting point for the calculation.

We do not know what really happened at the moment of the Earth's relative motion to the Sun that resulted in the current orbit of the Earth around it. The chance for the Earth to come to its current position by random explosion would be very low or close to zero. In Hugh Ross's book, *Improbable Planet*," he states the following.

> What several decades of research have revealed about Earth's location within the vastness of the cosmos can be summed up in this statement: the ideal place for any kind of life as we know it turns out to be a solar system like ours, within a galaxy like the Milky Way, within a supercluster of galaxies like the Virgo supercluster, within a super-supercluster like the Laniakea super-supercluster. In other words, we happen to live in the best, perhaps the one and only, neighborhood

that allows not only for physical life's existence but also for its enduring survival.[16]

On the other hand, the chance for the Earth to be at its current position through divine intervention is much higher, as stated by Newton in *The Principia*. Beside the astronomical and physical factors, the geological, chemical, biological and environmental processes are all important ingredients to be combined accurately to form human life. Too many unseen things with invisible qualities are necessary to make the observable things exist as they are now.

Y.C. Ruan used mathematical and physical principals to identify the boundary between life and natural phenomena by natural laws (L) and chance (C) and to prove the probability of life's origin by LC to be exactly zero.[17]

Only God with divine wisdom so great is capable of making life possible.

---

[16] Hugh Ross, Improbable planet: how earth became humanity's home, Grand Rapids, Michigan: Baker Books, 2016: 28

[17] Chu Dao(Y C Ruan), Life Made in Wisdom: the Mathematical Principles of Biointelligemce and the Origin of Life, Xulon Press, January 17, 2018

Chapter Five

LAW OF PHYSICS

I love mathematics because the wisdom behind arithmetic, algebra, geometry, calculus, and so on can be apprehended through computation, verification, and simulation. To further appreciate the wisdom of mathematics, I find applying it to physics fascinating. For instance, Newton's second law of motion using a mathematical formula to relate the force (F) to the mass (m) and acceleration (a).

Force (Sum of Force) = Mass X Acceleration.

Or, F = m * a

Under the correct unit platform, people can use this formula to determine the moving conditions of every object's motion (with speed V_0 much less than the speed of light V_c).

$V_0 < V_c$

When a force is applied to an object under the gravitational force, where a = g (g is the gravitational acceleration ~ 9.8m/s² on Earth), people can determine how much force is needed for a rocket to overcome the pull of gravity so it can fly to the outer atmosphere of the Earth. The equation combined with the law of gravity discovered by Newton is:

$$F = \frac{m_1 m_2}{r^2} G$$

The force of gravity between two bodies is proportional to the masses (m_1, m_2) and inversely proportional to the square of their separation distance (r). The force of gravity is always attractive for the masses in general. G is the gravitational constant = 6.67 X 10^{-11} N.m²/kg².

People can determine the exact time and location for the relative motion of the Earth to the moon and sun. A few hundred years ago, the king of the Ching dynasty in China was interested in comparing the forecast of moon eclipses from the Chinese traditional method versus the Western scientific method. Chinese astronomers have accumulated a few thousand years of experience and hard work to come up with the prediction of moon eclipses. It could give a rough estimation of when the moon's eclipse would occur based on years of experience and old data gathering. However, Western scientists applied the knowledge of Newton's second law, other related astronomical measurements, and laws accumulated in a relative short period of time, they could project almost to the exact minute and second when the moon's eclipse would start and end. The comparison demonstrated how important it is to know the laws of physics to help people understand the relationships among objects. Their application of it could benefit those people who master the knowledge behind the mathematics and laws of physics.

As science advanced into the twenty-first century, three schools of thought regarding God emerged.

1) Science replaces religion or God (Atheism)
   a. The more scientists know about the universe and physics law, the more the equation of God needed for the creation can be taken away or reduced.
   b. Even the universe had a beginning. There are different kinds of speculations and theories stating that the universe began without God.
   c. Evolution and the natural selection theories can explain where life and man came from without taking God into consideration.
2) Science confirms God's existence. (Theism)
   a. Science showed evidence for the origin of the universe through the "Big Bang Theory" confirms there is God.
   b. Information (analogy to a software program) stored in DNA points to God or the existence of an intelligent supreme being.
3) It is impossible to prove either God's presence or God's absence (Agnostics)
   a. People believe nothing is known or can be known of the existence of God, or anything beyond material phenomena.
   b. People are skeptical about the existence of God and claim neither faith nor disbelief in God.

I have read and watched arguments and debates about the existence of God. There are smart and intelligent people with very good training in science who choose their views or philosophies to embody any of the above schools of thought regarding the existence of God. The following analyses are my attempt to establish some common ground in preparation for future dialogue among scientists with different opinions:

A) Science can help screen or rule out if there are false beliefs. For example, the faith of the people who worship a "Monkey King[18]" in a temple will not be able to stand the trials or test of science. Investigating the related historical evidence relative to scientific checks, one can conclude that people who believe in a Monkey King are mainly superstitious.
B) Science needs evidence as a basis to establish any law or proposed theory, and these need to be subjected to verification and validation.
C) There are limitations in science, particularly related to the prediction of origin based on the projections back in time or in the future. For example, Newton's classical model of the universe assumes a steady universe, but it is not able to predict the origin of its beginning. Since the development of Albert Einstein's general relativity and the Big Bang theory in the last century, there is strong scientific evidence for the universe to have a beginning in time.

Seventy-seven scientists, who proclaimed their believe in the Bible and the God of the Bible, were listed in the *Institute for Creation Research*'s website.[19] They lived between fifteen to twenty centuries A.D. with contributions that laid the scientific foundation of modern sciences today. Had it not been their endeavor to discover the wonder of the universe, which they firmly attributed to the Creator's mighty power as revealed in the Bible, modern science and technologies would not be advanced to current achievements.

The next chapter, "Universal Order," will establish some common explanations of communications that lead to roadmaps of key ingredients for communication. Using these related roadmaps for chapter 13, "The Wisdom Communicated Through Special Revelation: Jesus Christ," I will present a picture of creation and communication via divine wisdom.

---

[18] Monkey King appeared as a main character in the 16th century Chinese classical novel Journey to the West (西游记) and is found in many later stories and adaptations. He is a monkey born from a stone who acquires supernatural powers through Taoist practices. (June, 2019 Ref.: https://en.wikipedia.org/wiki/Sun_Wukong)

[19] https://www.icr.org/article/bible-believing-scientists-past, accessed Oct 23, 2019

Chapter Six

UNIVERSAL ORDER

W hen you use a computer, laptop, or cell phone, the software and hardware have to work together as one unit to function properly and intelligently. The computer without software is just an assembly of materials made up of metal, plastic, semiconductor, and so on. On the other hand, the software of a computer needs a compatible and functionable hardware so the computer can do what it was originally designed to do. Even though the operation of the software is invisible to the user, it controls, organizes, and coordinates how the computer serves and interfaces with the users. Companies like Microsoft employ a lot of programmers to write, design, and apply the software codes for their many software programs. The intelligent order of operations and related information on any computer screen can be traced back to the companies' programmers.

Compared to the computer programs, biological information (DNA) is the most important information humans have ever discovered. The founder of Microsoft, Bill Gates, wrote in a book, "Human DNA is like a computer program, but far, far more advanced than any software we've ever created."[20]

The human genome contains approximately three billion pairs of DNA, which reside in the twenty-three pairs of chromosomes within the nucleus of all human beings' cells. Each chromosome contains hundreds to thousands of genes, which carry the instructions for making proteins. A genome is an organism's complete set of deoxyribonucleic acid (DNA), a chemical compound that contains the genetic instructions needed to develop and direct the activities of every organism.

DNA molecules are made of two twisting, paired strands. Each strand is made of four chemical units, called nucleotide bases. The bases are adenine

---

[20] Bill Gates, The Road Ahead, Penguin: London, Revised 1996: 228

(A), thymine (T), guanine (G), and cytosine (C). Bases on opposite strands pair specifically: an A always pairs with a T, and a C always with a G.[21]

Biologists have been gradually learning that the basic cellular unit underlying all known life on Earth is extremely complex and compact. A human being is comprised of trillions of cells[22], different types of proteins, organs, muscles, bones, nerves, etc. The average diameter of a cell nucleus, where DNA information is stored, is approximately six micrometers (μm), which occupies about ten percent of the total cell volume.[23] The storage and functional capabilities of the cell nucleus, with respect to its size, are much superior to current electronic technologies. For example, a typical DVD holds 4.7 GB of data while a cellular nucleus holds about three billion data (or approximately 3 GB). However, the size and the complexity of the DNA is more compact and better designed.

We understand without hesitation that electronic products like computers, robots, cell phones, and so on, require human designers to develop and implement them for use in certain applications. Even the latest development of artificial intelligence (AI) can perform some learning and problem solving. We know that AI is developed by human beings to mimic certain functions routinely done by the human mind. If electronic products are made by an intelligent being, the human being himself can be traced back to being made by a supreme being we call God.

Bill Gates, founder of Microsoft, was asked if he believed in God. "The mystery and the beauty of the world is overwhelmingly amazing, and there's no scientific explanation of how it came about. To say that it was generated by random numbers, that does seem, you know, sort of an uncharitable view. I think it makes sense to believe in God,"[24]

If you wonder where you and your ancestors came from, there are generally two basic explanations. First, a supernatural being created the universe and human beings. Alternatively, the entire universe, including human beings, somehow evolved from natural selection or survival of the fittest. Following this latter argument, the universe came from a steady state, a

---

[21] https://www.genome.gov/human-genome-project/Completion-FAQ

[22] Yella Hewings-Martin, How many cells are in the human body? 12 July 2017, https://www.medicalnewstoday.com/articles/318342.php

[23] https://en.m.wikipedia.org/wiki/Cell_nucleus, accessed June, 2019

[24] https://www.rollingstone.com/culture/culture-news/bill-gates-the-rolling-stone-interview-111915/#ixzz30rCahWUD, March 27th, 2014

self-existing universe without a beginning. This hypothesis contradicts natural laws such as thermodynamics, which I will further discuss in the next chapter, "In the Beginning." Also, people argue from this viewpoint that the universe had a beginning and then randomly developed together with natural selection, which resulted in this universe and human beings. This explanation would be unlikely from the physics point of view.[25] The law of entropy states that physical processes result in more disorder and increase randomness unless external organization comes in to improve the process. Therefore, it is logical to explain the origin of the universe and life on Earth as caused by a supernatural being, or a Creator. People can find evidence everywhere if they really pay attention to the visible and invisible qualities in heaven and Earth.

Physical objects in our visible world are first used to illustrate that random processes cannot produce orderly products. If you put some sticks, wood, knives, screws and nuts inside an enclosed container, they will not join together to form a meaningful object or product, even after you randomly shake and move the container a trillion times.

Another example relates to a watch picked up by a person in a desert. He or she will no doubt know right away that someone designed the watch based on its mechanism and appearance. No one would assume it came from evolution because of the windy and sandy environment, or that it was assembled out of random processes. Human beings are much more complicated than a watch, so the analogy of a Designer who designed us also make sense.

People may argue that on a microscopic scale or in an invisible world, meaningful things may be formed, given enough time, according to quantum fluctuation and processes. Let us look at some of the random quantum systems with dimensions in the order of DNA[26] (diameter ~ 2nm), which are inside a human cellular nucleus.

My doctoral thesis relates to the study of random systems in comparison with periodic and quasi-periodic systems.[27] The investigation used computer simulation and experimental results to find the outcome of different types

---

[25] https://www.creation-evidence.com/

[26] DNA molecule is 34 angstroms long and 21 angstroms wide: a golden ratio, http://www.math.brown.edu/~banchoff/ups/group5/ma8n_animals2.0.html

[27] Alan Chi-Chung Tai, Study of unconfined states in quasi-periodic semiconductor super-lattices, Ph.D. Thesis, Boston College, 1991, University Microfilms International, Order Number or ProQuest publication number: 9211799 (https://dissexpress.proquest.com)

of quantum wells with a dimension of 2.8 x 10 $^{-9}$m (2.8nm). The physical part of the quantum well is a superlattice semiconductor, but the analytical approaches of the studies can be applied and extended to other structures. The quantum wells were generated by very thin and uniform layers of two types of semiconductors: (AlGaAs as a Quantum barrier corresponding to "1" and GaAs as a quantum well corresponding to "0") according to certain mathematical tables.

For example, periodic structure periodically arranged itself like 1010101010101... while random structure randomly arranged itself like 1111001000001011... Fibonacci structure formed into a Fibonacci arrangement of 01011010110110101101011... where the third group of sequences was formed by the two former groups of sequences combined.

The experiments showed that the random structures did not form a well-defined energy level within the well, nor above the well (unconfined state). The periodic structures had a well-defined energy level in two dimensions (2D), where electrons were confined to move in 2D because of the quantum mechanical characteristic within the quantum wells, inside the well, but not in the unconfined state in three dimensions (3D). The Fibonacci structures had a well-defined energy level in 2D within the wells. The interesting results happened in the unconfined states, where the electrons could move in three dimensions and still had a well-defined energy level. Computer simulations were used to check different types of quasiperiodic sequences, which I designed intentionally to follow certain mathematical equations. Similar experimental results were observed, as the Fibonacci-sequenced structures demonstrated. The results also agreed with the simulation prediction about what would happen in the quantum nature on a microscopic level.

The conclusion shows Mother Nature does not favor random composition and processes in generating meaningful results and outcome even on the microscopic scale. The introduction of mathematical equations on these experiments successfully confirmed that a better outcome of physics and science could be designed. In this case, the designer was me.

Consequently, the amazingly sophisticated arrangement and order of this universe could only be possible by pointing to a Creator, who was behind all the well-designed formulation of all visible and invisible things operating within this universe. In addition, He is also a Designer above and beyond the material things of this universe. As clearly communicated through the

Bible, both Old Testament[28] and New Testament: "God is Spirit, and those who worship Him must worship in spirit and truth."[29] To my understanding of God's Spirit, He is not limited by time and space. Therefore, He is above and beyond the material things of the universe.

The beautiful world we see today is mainly comprised of periodic and quasi-periodic materials, reflecting the Designer who designed them with wisdom and special mathematical rules. One of the examples for the quasi-periodic sequence is the Fibonacci sequence. The number of petals on a flower, for instance, will often be a Fibonacci number. The seeds of sunflowers and pinecones twist in opposing spirals of Fibonacci numbers. The "golden ratio" is best approximated by Fibonacci numbers. This ratio is sometimes called the "divine proportion," because of its frequency in the natural world.[30] The golden ratio appeared in nature to make the world more beautiful and practical than ordinary periodic structures. The quasi-periodic crystal gives rise to fivefold symmetry in an x-ray diffraction pattern not possible in periodic crystals[31], nor in random-structure crystals.

All these appearances of the special, beautiful, and mathematical structures in nature boil down to the creative design of a Creator who, in human terms, is also a mathematician, a scientist, and an artist. He wants to communicate to all human beings about His glory through His creation and wisdom as recorded in Psalm 19:

> The heavens declare the glory of God;
> the skies proclaim the work of his hands.
> Day after day they pour forth speech;
> night after night they reveal knowledge.
> They have no speech, they use no words;
> no sound is heard from them.
> Yet their voice goes out into all the earth,
> their words to the ends of the world.[32]

---

[28] Genesis 1:2, "...and the Spirit of God was hovering over the waters."

[29] John 4:24, NKJV Bible

[30] https://www.nationalgeographic.org/media/golden-ratio/

[31] H.C. Von Baeyer, Discover, 69, Feb., 1990

[32] Psalm, 19:1–4

Mia Chung—one of America's highest ranking pianists, distinguished educators, and music advocates—is a professor of interpretive analysis at the Curtis Institute of Music. Professor Chung shared her thought about mathematics, music and Creator.

> For centuries, scholars and mathematicians viewed the math of the universe, whether found in music or the petals of a flower, as a manifestation of God's creative genius. The Bible says that man was 'created in the image of God.' Nowhere is this image more evident than in the human capacity to compose, invent and create. But if we are created in the image of the Creator of the universe, is it possible that the Creator of the universe is reaching out to us through math and music and our own creative potential? The math in music speaks to the heart of a creative God who wants to communicate to us. He does so not only through the natural laws, but also through our emotions and spirit. He aims to speak directly to us.[33]

There are also universal orders in communications, which I outline below based on my observation and training in physics and science. In general, communication involves at least two parties, involving hardware, software, and messaging.

For example, one computer and one cell phone are communicating in different locations. Hardware is set up functionally to be compatible for communication between the computer and a cell phone. The media connecting them can be wired or wireless, but the context or message transmitted from one end and received on the other end needs to be accurately connected and contacted.

I used the triune T shape to represent the basic ingredients for communication in current technical terms. In Figure 1, there is the "Source" of original representation on the left, the "Expression" as the relational representation on the right, and the "Connection" as the messaging representation. These can also be identified as a triune expression like a triangle (one unit with three sides), but expressed in a T shape. The term triune in this case means only the three embedded characteristics of the one unit of an entity—the computer as an example. The function of this special expression

---

[33] Mia Chung, http://www.veritas.org/missingpieces/

will be explained in the following science and life section with some examples related to communication or building certain types of relationships.

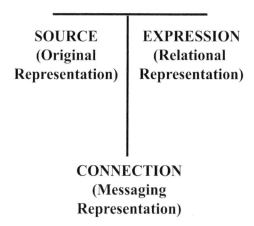

SOURCE | **EXPRESSION**
(Original | **(Relational**
Representation) | **Representation)**

CONNECTION
(Messaging
Representation)

Figure 1: triune T expression

In the simplified diagram of Figure 2, Computer 1 on the left of "Communication T" is an individual entity with a general term described as "Computer." In this same T, Cell phone 2 is an individual entity with a general term described as "Cell phone." The communication T on the left side only has two parts that show what the communication parts are in the beginning. The top expression of "triune T1" with software, hardware, and message are based on Computer 1. This T-shaped expression facilitates the three arrangements: invisible source as software, physical expression as hardware, and media connections with the content of messages. As in Figure 1, the entity that will communicate with another entity, in general composed of three parts, depends on what type of communication will be established.

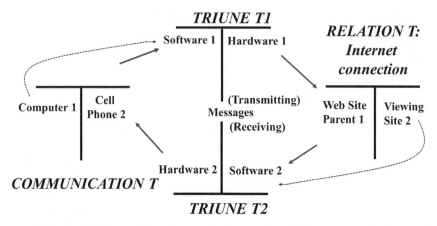

Figure 2: Example of computer and cell phone roadmap

These triune arrangements in Figure 2 can be treated as one whole unit as the representations of Computer 1. On the right side of the diagram, the "Relational T," with a T-shaped letter, umbrellas two items interrelated in a specific relationship. In this example, the item on the left of the "Relational T" is a specific parent website that comes directly from Computer 1 with all the needed preparation, intelligence, and coordination. The item next to it on the right side is a specific reader who can view and respond to the parent website. Therefore, this particular example demonstrates the relationship of a virtual parent website with a virtual child user. To establish this relationship, the lower diagram with an upside-down T-shaped letter represents the triune reflection, or image of Computer 1 in the upper diagram. It can be an Internet cell phone made up of software (invisible source), hardware (physical configuration) to interact within the Internet connection, and the message connection through the same media for communication as Computer 1.

The arrows in this picture start clockwise from Computer 1 in the top left corner to connect the related item. Eventually, the arrow passes through the "relational" T-shape and will end where it starts with the completion of "Cell phone 2." The complete cycle in this case presents the relationship of two entities that can be connected and unified through a specific relationship. Cell phone 2 is marked as compatible with both hardware and software to interpret and decode the meaning of the message transferred to it. This allows the user of the cell phone to understand the message and respond to the computer at the proper time.

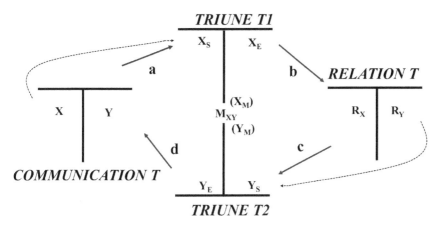

Figure 3: Generalized roadmap

We can generalize this relationship diagram by joining X and Y with respect to their relations in Figure 3. It starts with Communication T, with X represented by the triune T1: X_S as the source, X_E as the expression, and X_M as the message to be transmitted and received. Relation T with R_X and R_Y correspond to the relationship of X and Y.

Using the same principle, T2 is the triune function of Y: Y_S as the source, Y_E as the expression, and Y_M as the message to be connected and unified to X_M. After the unification and connection, X_M and Y_M are viewed to be combined and are represented by the oneness symbol of M_{XY}.

These relational cycles are joined together with the processing arrows "a" to "d," moved and joined through the communication T, triune T1, relation T, triune T2 and then back to communication T in a clockwise direction. The picture gives us a scientific map showing the relationship between X and Y, and the processes of communication media. Y is basically related to X through the establishment of Rx and Ry by passing the message of M_{XY} that is united functionally through the triune setting.

The modeling of the map in communications appears simple, but can be profound in the science of communication, so the related wisdom can be appreciated by human beings. There will be more examples demonstrating how to apply this principle for relational and communication studies. For easy reference, this roadmap of communication uses different T-shaped letters with respect to their compositions and relationships, I call this map "T transform."[34]

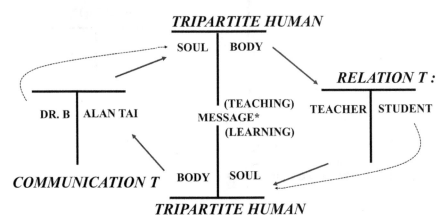

Figure 4: Teacher and student roadmap
*Note: The relationship between teacher and student mainly allows the teacher to communicate with the students intelligential information/messages associated with the soul.

The map in Figure 4 for the relation between a teacher and a student can be established in similar fashion. For illustrative purposes, I'll name the teacher Dr. B and the student Alan Tai to show this example of communication between two individuals. Dr. B is a human being in the role of teacher, who will teach another person, Alan Tai, in the role of student. During any of the teaching processes under this relationship between teacher and student, Dr. B uses his soul (invisible source) to think and organize his teaching information. He passes it to his body (physical and visible part of Dr. B) to be expressed in sound (speaking language) and sight (body language), etc. The sound and sight need to be meaningful, containing messages and content that can be conveyed and understood by the student. There could have been other means of one-way or two-way communication

---

[34] To the best of my knowledge, there is not yet usage of this "T transform" for the presentation of communication and relation connections. There will be more examples for the applications of the T transform to be used in science and life in this book and my future publications. The purpose is to facilitate the interpretation and transformation in one way, two ways, or more ways of communication. It can be applied to certain universal orders: physics, math, and Biblical relations, etc. T transform could also be called "Tai's Transform," but I leave that to future feedback if the general public will find these types of transform to be useful and acceptable in different areas of application.

in this T transform, including emotional and will communication between the teacher and student.

In general, the human soul is comprised of at least mind, emotion, and will, so corresponding communication can be transformed and passed through the interactions of the two or more parties. In this example, the parties are teacher and student who form the relationship and communication. Basically, teacher and student relations are established through the communication of relevant information.

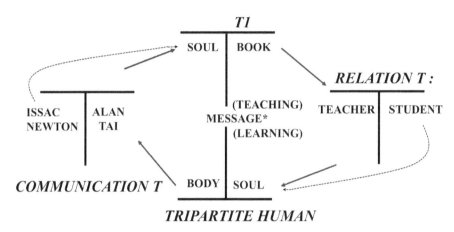

Figure 5: Teacher and student roadmap 2

Even a teacher who lived in the past can leave behind his or her teaching, philosophy, insight, and life principles. Any human interested in learning from such a "teacher," through reading and studying his or her related information, can become his or her student. Take the example of Isaac Newton, whom I highly respect and always want to learn from. I can be his "student," as shown in Figure 5.

The Expression part in T1 is a book instead of a living being. I read his biographies, used the Newton laws, and even visited Cambridge University, where he was a teacher/professor. I read his original handwritten notes in a book displayed inside the library. Over the years, I "interacted" with him through the information I gathered so I became his virtual student.

The illustration of teacher and student relationships nowadays can be seen in my studying for the Graduate Certificate of Biblical Foundations at Grand Canyon University. Using the latest technologies for online classes, I was able to communicate with a professor called Dr. B in two classes. I never even physically met him, but he was able provide excellent teaching, sharing, questioning, and coaching in the classes of "Biblical Hermeneutics" and "Christian Character Formation."

As a virtual student, to meet the requirement of the classes, I needed to read and do all the assignments and discussions in the online forum with other students. However, the most important lessons I learned were through Dr. B's coaching and his frequent involvement and sharing in discussion groups using real-life examples. I also talked to him on the telephone to receive guidance and coaching for research papers.

The difference between being a "virtual student" of Isaac Newton or of Dr. B was the interaction as illustrated in the T transform of one-way and two-way communication. However, both teachers were able to transform my soul of learning in science and life, even though Newton is no longer alive. This teacher-student relationship, along with their wisdom, will be treasured and appreciated by not only me but all those willing to learn from the teachers.

In theoretical physics, Feynman diagrams are pictorial representations of the mathematical expressions[35] describing the behavior of subatomic particles. The interactions of subatomic particles can be complex and difficult to understand intuitively. Feynman diagrams give a simple picture for people to visualize. The basic concept of the relationship among the particle's interaction is out of the scope of this book, so I won't go into the detail of the mathematical and physical model related to science of Feynman diagrams. I just used a simple illustration of the interaction of two electrons that results in the electromagnetic repulsive force. Recent advances in science helped scientists identify these fundamental forces as either an electromagnetic force, a strong force, a weak force, or a gravitational force. The four fundamental forces of the universe indirectly sustain all material, planets, stars, and galaxies in current expression and movement.

The application of the electromagnetic force to human beings in modern science impacts all walks of life. From electric toothbrushes to automobiles, everything that uses an electric motor can reduce human work, and acts indirectly as a laborer to be controlled by humans. All these became possible when scientists discovered electrons and came to understand the physics behind electromagnetic forces. As in Figure 6, the typical Feynman diagram for the two-electron interaction shows the electrons exchanging a "virtual photon" between their initial states and final states.

---

[35] Feynman diagram, Wikipedia, https://en.wikipedia.org/wiki/ Feynman diagram, accessed June 17, 2019

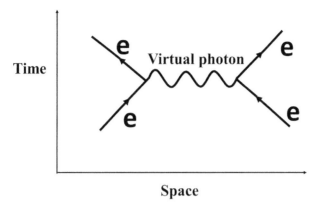

Figure 6: Feynman diagram for electromagnetic force between electrons

The X axis corresponds to space and the Y axis corresponds to time. Which axis, may not be indicated sometimes in the presentation. The Feynman diagram can also be rotated 90 degrees so the X axis corresponds to time while the Y axis corresponds to space. This principle of repelling force from the exchange of virtual photons can be illustrated as a ball thrown between two parties. One party throws the ball while the other party receives the ball shortly after the pushing processes.

Newton's third law shows that action and reaction are equal, but opposite. Therefore, the two parties get a repelling force of pushing them away from each other. This process repeats continually as the two electrons pass through each other, so the electromagnetic forces would push the two electrons away from each other. This principle of electrons' interaction is then rearranged to put into the T transform diagram in Figure 7.

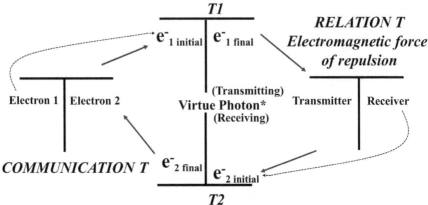

Figure 7: Electromagnetic force roadmap

*Note: Reference to Richard Feynman's diagram that showed a graphical representation of particle interactions**

This universal order in physics has the potential to be studied systematically and used together mathematically in understanding science and life. Without going into too much detail, the appendix of this book will have more general discussions about interesting mathematic investigations using the T transform, which I plan to publish in mathematics or physics journals.

I also see, intuitively, some kind of universal order for the information in the DNA sequence. This is out of the scope of my background of studies and training in science. I hope some researchers will investigate this DNA sequence by mathematical analysis using the T transform or other information-studying tool.

The mystery of life can be appreciated greatly when scientists understand divine wisdom lies behind all amazing creation. I am definitely impressed by the creative power and wisdom of the Creator, who created the universe "in the beginning" and upholds the universe by His mighty power.

THE WISDOM COMMUNICATED
THROUGH ETERNAL POWER

Chapter Seven

IN THE BEGINNING

66 "In the beginning God created the heavens and the earth. Now the earth was formless and empty, darkness was over the surface of the deep, and the Spirit of God was hovering over the waters. And God said, "Let there be light," and there was light.[36]"

What an amazing statement from God revealing His work as the owner and Creator of the heavens and Earth! There are not many details explaining how He created the universe, other than His creation followed His exact Word.

The three sentences reveal to humans several important revelations about God and the universe:

1) There was a beginning in time in the universe.
2) God existed as He was before the beginning of time.
3) God has the power to create materials things.
4) Before the beginning of time, there were no material things (nothing) apart from God.
5) The Earth existed at first in chaos.
6) The Spirit of God was preparing[37] to bring order.
7) When God said there should be light in a certain time and space, light shone.

Genesis was written more than three thousand years ago with a unique revelation that introduced the beginning of the universe. God is Spirit[38] and is not bounded by time and space. God is beyond the material, so all

---

[36] Genesis 1:1–3

[37] George Montague, https://wau.org/resources/article/re_he_still_hovers_over_chaos/, accessed June 19, 2019; "Because God's spirit was hovering over it, chaos became promise and order."

[38] Genesis 1:2; John 4:24

things can be created out of nothing[39]. The first two chapters also tell us the only God, who created humans in the image of God, wanted humans to rule the Earth[40]. This idea contrasts with the Middle Eastern religions[41] during the time Genesis was written. The religions in the regions are polytheistic and have different types and ranking of gods. They believe gods created humans to be their slaves. Their gods also possess many similarities with humans that can be traced back to origins in the material realm. However, as discussed below, material things would not be manifested yet at the beginning of time.

The scientific evidence of the universe's beginning started with the experimental data provided by Edwin Hubble in 1929.[42] Using the Doppler effect, he confirmed the universe was expanding. This well-known physics law is related to the frequency shift of light, sound, or other waves and the relative velocity of a moving object—like the changing sound of a firetruck siren as it drives past. In this case, the moving objects were the galaxies emitting their light. His analysis showed the farther away the galaxies were from the Earth, the faster they moved away from it. When the physical and theoretical model of the universe was generated based on the scientific data, the backtracking led to a belief that the universe had a beginning.

Francis Collins wrote in *The Language of God* about the Big Bang theory: "Based on these and other observations, physicists are in agreement that the universe began as an infinitely dense, dimensionless point of pure energy. The law of physics breaks down in this circumstance, referred to as a 'singularity.' At least so far, scientists have been unable to interpret the very earliest events in the explosion, occupying the first 10^{-43} sec. After that, it is possible to make predictions about the event that would need to have occurred to result in today's observable universe..."[43] Before the Big Bang theory was verified, scientists disagreed on how the universe started and a lot of them, including Albert Einstein and others, favored a

---

[39] Ian Hutchinson, Can a scientist believe in miracles? InterVarsity Press, 2018: 119, the universe had a beginning and was created by God ex nihilo, "out of nothing"

[40] Genesis 1:27–28

[41] Ancient Mesopotamian religion, Wikipedia, https://en.wikipedia.org/wiki/Ancient_Mesopotamian_religion, accessed June 19, 2019

[42] Francis Collins, The Language of God: A Scientist Presents Evidence for Belief, Simon & Schuster, 2006: 63

[43] Ibid, 65

steady universe model. The Christians, including those scientists trusting the Bible as the Word of God, already believed the universe had a beginning. The introduction of the Bible in Geneses 1:1 clearly stated that centuries ago. Only the Creator, who is beyond time and space, can communicate to humans with such accuracy and authority. One may argue a fifty percent chance for the truth of a statement that a beginning in the universe was predicted by some wise human being. However, when you combine the fifty percent chance of multiple statements that agree with scientific discoveries, the chance grows very small that humans from generations ago predicted this as true.

For example, the chance of continuing to be correct using ten forecasts with a fifty percent chance of accuracy each is $(0.5)^{10}$, or $1/1024$. Whereas, for fifty forecasts the chance for accuracy is $(0.5)^{50}$, or $1/1.1259 \times 10^{15}$. On the other hand, even more remarkable, no Biblical statements conflict with the latest scientific discoveries. To clarify, the Bible's introduction did not start with a statement that there was no beginning in heaven and Earth, nor that the Earth was flat.

Some of the Biblical statements not quite understood now will, one day, be found in agreement when science advances enough. Of course, this statement is my faith-based projection of what will come true based on the existing strong scientific evidence we have in comparing Biblical statements. Only a God beyond time and space can reveal the Truth regarding what humans need to know, as in the following Biblical statement.

"All Scripture is God-breathed and is useful for teaching, rebuking, correcting and training in righteousness, so that the servant of God may be thoroughly equipped for every good work."[44]

John Lennox, a Christian mathematician, shared the shocking experiences of a nonbeliever's point of view after comparing scientific discoveries with the subsequent creation events in Genesis 1.[45] Lennox used the example of an atheist scientist, Andrew Parker, a research director at the Natural History Museum in London, to show feedback from people who did not profess to believe in God. Parker, an evolutionary biologist, was stimulated to look at Genesis 1 after a number of people wrote to him suggesting his research on the origin of the eye seemed to echo the statement "let there be light."

---

[44] II Tim 3:16–17

[45] John C. Lennox, Seven days that divide the world: the beginning according to Genesis and science, Zondervan, 2011. Appendix B, the cosmic temple view

"I discovered a whole series of parallels between the creation story on the Bible's first page and the modern, scientific account of life's history. This at least made me think. The congruence was almost exact." He later adds, "The more detail is examined, the more convincing and remarkable I believe the parallels become. One question I will be asking in this book is this: could it be that the creation account on page one of Genesis was written as it is because that is how the sequence of events really happened?"[46]

Parker concluded with, "Here, then, is the Genesis Enigma: The opening page of Genesis is scientifically accurate but was written long before the science was known. How did the writer of this page come to write this creation account? ... I must admit, rather nervously as a scientist averse to entertaining such an idea, that the evidence that the writer of the opening page of the Bible was divinely inspired is the product of divine inspiration is strong. I have never before encountered such powerful impartial evidence that the Bible is the product of divine inspiration."[47]

Numerous books discuss the creation according to the Bible and modern science. Some of them are listed in the reference section of this book for further exploration. This reflection from Nathan Aviezer sums up his insights: "There is consistency between contemporary scientific knowledge and the literal meaning of the first chapter of the Book of Genesis... Nevertheless, it is inevitable that some parts of the analysis will require revision as additional scientific data becomes available. However, we expect the basic pattern to remain intact. In fact, we anticipate that the scientific knowledge of the future will lead to additional understanding and new insights into those biblical passages that are still unclear."[48]

I am most interested in sharing insights into communication through divine wisdom that leads to our appreciation of the wisdom in science and life. Genesis 1:27 tells us humans were created "in the image of God" on the sixth day. We humans, as opposed to animals, also have the unique capability to communicate with others much more intellectually. For example: human civilization started when written words were used to communicate to one another, while the Information Age started when the Internet was used to communicate more effectively.

---

[46] Andrew Parker, The Genesis Enigma (London: Doubleday, 2009): xiixiii.

[47] Ibid, 238

[48] Nathan Aviezer, In the beginning: Biblical creation and science, Hoboken, N.J.: Ktav Pub. House, 1990: 121

Nathan Aviezer pointed out, "The past few thousand years have witnessed the enormous progress made by man in all areas of intellectual endeavor. An essential ingredient of this progress is the unique ability of members of the human species to communicate with one another. This enables man to benefit from the accomplishments of his predecessors. The distinguished physicist Isaac Newton once remarked: 'If I have seen further [than others], it is by standing upon the shoulders of giants.'"[49]

Various commentators explain that Biblical days in Genesis mean a phase or a period in the development of the world. [50] There are different viewpoints around this topic, but John Lennox[51] pointed out one interesting insight that also resonated with me. I summarized it, adding my own understanding of physics:

1) When God spoke His Word, special events — described as "singularities"[52] — happened.
2) The creation of heaven and Earth gave rise to physics laws that followed God's speaking.
3) During the period of time when God's Word was not mentioned, the laws of physics governed all materials of creation, while God's Word still upheld and maintained the operation of the universe.[53]
4) Between the days of the creation, there was "And God said..." which could imply there were millions of years when things and living things followed the law of physics until the next event.
5) God spoke twice in day three and six, while the other days mentioned that God spoke only once. In day three, "then God said"[54] could imply God turned the singularity event from inorganic to

---

[49] Ibid, 110

[50] Ibid, 1–2

[51] John C. Lennox, Seven days that divide the world: the beginning according to Genesis and science, Zondervan, 2011.

[52] Singularity is used mostly in mathematics to express a point that is not well defined like infinity. In comparing with the Word of God "Logos" gives rise to the law of physics not limited by time and space, the Word of God "Rhema" gives rise to a specific event (singularity) that happened at certain time and space. (my interpretation)

[53] Hebrews 1:3a, "The Son is the radiance of God's glory and the exact representation of his being, sustaining all things by his powerful word."

[54] Genesis 1:11

organic living things (lifeless to life) in preparation for the next biggest event of day six.

6) On day six came the climax of all creation mentioned after "then God said:"[55] "Let us make mankind in our image, in our likeness, so that they may rule over the fish in the sea and the birds in the sky, over the livestock and all the wild animals, and over all the creatures that move along the ground." God manifested the singularity event which added humans on Earth, the only living beings with spirit and soul. This concluded His creation in six days.

7) By the seventh day, God had finished his work. He blessed the seventh day and made it holy, because He rested from His work of creation.[56] Again, this does not mean God left the creation and rested forever. After the seventh day, God still spoke directly or indirectly on Earth and in heaven. His works and words are continually being witnessed by His people and His son Jesus Christ.[57]

How great this universe is and how special the Earth and human beings are. Psalm 104 shows some of the marvelous works the Creator has done:

How many are your works, Lord!
In wisdom you made them all;
the earth is full of your creatures.
There is the sea, vast and spacious,
teeming with creatures beyond number—
living things both large and small.
There the ships go to and fro,
and Leviathan, which you formed to frolic there.
All creatures look to you
to give them their food at the proper time.
When you give it to them,
they gather it up;
when you open your hand,
they are satisfied with good things.

---

[55] Genesis 1:26

[56] Genesis 2:2–3

[57] John 5:17, "In his defense Jesus said to them, "My Father is always at his work to this very day, and I too am working."

When you hide your face,
they are terrified;
when you take away their breath,
they die and return to the dust.
When you send your Spirit,
they are created,
and you renew the face of the ground....[58]

---

[58] Psalm 104:24–30

Chapter Eight

THE VASTNESS OF THE UNIVERSE

The vastness of this universe and the uniqueness of the Earth point us to ask questions about why we exist in this particular space and time. What is the purpose of our life on this Earth? Is there a Creator behind this amazing universe and our world? How can we know and communicate with the Creator? So many of these questions have been asked over the ages in human civilization.

A review of the size of the Milky Way gives people some idea the vastness of the galaxy we live in. The latest scientific measurement of the Milky Way's diameter is about 100,000 light years[59]. This vague number usually provides little understanding of the vastness of its size, even though we know the distance of one light year is the distance for light to travel in one year. That's why researchers like to use different scales to compare the size of the solar system. The below examples are used by the Center for Astrophysics (CFA) at Harvard University to help people make a mental model in their mind about the size of the Milky Way and the universe:

1) Reducing the scale of our entire solar system to the size of a quarter gives us perspective about the size of Milky Way with respect to our solar system[60]. On this scale, the overall size of our Milky Way galaxy would be close to the size of the United States! Since the sun is at the center but smaller than the solar system, the sun appears like a small piece of dust based on the model scale of Milky Way to the United States. If we shrink the Milky Way galaxy to the scale of 1 m², the entire solar system is smaller than the nucleus of a human

---

[59] https://www.cfa.harvard.edu/seuforum/howfar/across.html, accessed June 22, 2019

[60] https://www.cfa.harvard.edu/seuforum/howfar/howfar.html, accessed June 22, 2019

cell. The Earth is even much smaller in this scale[61], let alone the size of humans with respect to our galaxy.

2) Another helpful mental model shows the comparative size of the universe we can see. Our entire Milky Way galaxy is further reduced to the scale size of a CD. The adjacent galaxy that is closest to Milky Way is the spiral galaxy, name "Andromeda." On this scale, Andromeda will appear to be about eight feet away. Due to the time it takes light to travel in space, the farthest galaxies we have ever seen once its light reaches us, as seen in the Hubble Deep Field, would be about nine miles away. When the light from the edge of the observable universe reaches us, it shows us the beginning of the universe during the start of the Big Bang. Therefore, it gives us some idea about the age of the universe[62]. The furthest we can possibly see of the entire universe is about ten miles from us on the scale of Milky Way to a regular CD. This corresponds to about fourteen billion light years away from our planet Earth.

How big is the universe?

No one knows how big the universe would be, as the light beyond the furthest and observable edge of the universe has yet to reach us. We can wonder if the universe is infinitely large, or if ours is the only universe out in the space. Astronomers have secondary evidence and some physical models to predict that the universe of galaxies spreads far beyond the region we can observe. How big this universe is and its billions of years of existence will blow our minds and imaginations!

The physical parameters of the observable universe can be estimated for its mass and size, assuming different types of materials fill it. The critical density of the universe is calculated to be $0.85 \times 10^{-26} \text{kg/m}^3$ (commonly quoted as about five hydrogen atoms per cubic meter)[63]. This density includes four significant types of energy/mass: ordinary matter (4.8%), neutrinos (0.1%),

---

[61] Reference to ratio of solar system diameter to earth diameter: Richard Feynman, The Feynman Lectures on Physics, Vol.1, Chaps.1, 2, and 3.

[62] Estimated as of 2015 around 13.799 ± 0.021 billion years for the age of the universe, Planck Collaboration (2016). "Planck 2015 results. XIII. Cosmological parameters (See Table 4 on page 32 of pdf)." Astronomy and Astrophysics. 594: A13.

[63] Observable universe, https://en.wikipedia.org/wiki/Observable_universe, accessed June 24, 2019

cold dark matter (26.8%), and dark energy (68.3%).[64] The density of ordinary matter, as measured by Planck, is 4.8% of the total critical density or 4.08×10^{-28}kg/m^3. To convert this density to mass we must multiply by volume, a value based on the radius of the "observable universe." Since the universe has been expanding for 13.8 billion years, the comoving distance of the cosmological scale is now projected to about 46.6 billion light-years for the radius of the universe. Thus, volume $(4\pi r^3/3)$ equals 3.58×10^{80} m^3 and the mass of ordinary matter equals density $(4.08\times10^{-28}$kg/m^3) times volume $(3.58\times10^{80}$ m^3) or 1.46×10^{53}kg. Note that the average human weight is about 60kg or 6×10^1kg. Therefore, the weight ratio of a human to the observable universe is $\sim 1/(2.43\times10^{51})$.

For a mental model of shrinking the observable universe down to the weight scale of the Earth (5.972×10^{24}kg), the weight of the human would appear to be in the order of a hydrogen atom (1.674×10^{-27}kg), which is incredibly small in a microscopic view.

These calculations and mental models present us with the fact that humans are extremely small. We are negligible in terms of size, relative to the space beyond our Earth. In summary, Richard Feynman in his famous Feynman lecture showed us the order of magnitude[65] to the heavenly objects using human size as 1 unit ($\sim$ 1m):

Diameter in m

Particle	$\sim10^{-15}$
Atom	$\sim10^{-12}$
DNA[66]	$\sim10^{-9}$
Cell	$\sim10^{-5}$
Human	~1
Earth	$\sim10^7$
Sun	$\sim10^9$
Solar system	$\sim10^{13}$

---

[64] Planck collaboration (2013). "Planck 2013 results. XVI. Cosmological parameters." Astronomy and Astrophysics. 571: A16

[65] Richard Feynman, The Feynman Lectures on Physics, Vol.1, Chaps.1, 2, and 3.

[66] New to Feynman Lectures, as DNA was not a common term when Feynman gave his lectures in California Institute of Technology (Caltech), during 1961–1963.

Milky Way	$\sim 10^{20}$
Visible Universe	$\sim 10^{26}$

The New York Times published a press release in 2016 stating the number of galaxies in the universe is at least two trillion.[67] The scale of the universe, already unfathomable, just became even more so: there are about ten times as many galaxies as previously thought. The new number, two trillion galaxies, is the result of work led by Christopher J. Conselice, an astrophysicist at the University of Nottingham in England, published in *The Astrophysical Journal*[68]. Our Milky Way solar system, where humans live, is only one of the two trillion galaxies!

On the other hand, there are trillions of cells inside each human being. Each cell stores about three billion pairs of DNA that hold information for the development of physical structures in human beings. The latest research about the universe and human body is still in its infancy as it relates to understanding science and life. The evolution theory proposed by Darwin and others could not explain how the information inside the DNA became available at the beginning. The astounding discoveries of the vastness of this universe and the complication of the biology/chemistry/physics inside a human being open our eyes to the mysterious and marvelous works created by the wisest Creator.

Psalm 8, written about three thousand years ago, tells us the universal and amazing response to the vastness of the heaven and Earth.

Lord, our Lord,
how majestic is your name in all the earth!
You have set your glory in the heavens.
Through the praise of children and infants
you have established a stronghold against your enemies,
to silence the foe and the avenger.
When I consider your heavens, the work of your fingers,
the moon and the stars, which you have set in place,
what is mankind that you are mindful of them,

---

[67] https://www.nytimes.com/2016/10/18/science/two-trillion-galaxies-at-the-very-least.html, accessed June 24, 2019

[68] Christopher J. Conselice, et al. (2016). "The Evolution of Galaxy Number Density at z < 8 and Its Implications." The Astrophysical Journal. 830 (2): 83.

human beings that you care for them?[69]

Chapter Nine

THE UNIQUENESS OF THE EARTH AND HUMANS

A mong trillions of galaxies and billions of stars in each of the galaxies, people may ask why the Earth's unique placement and structure allowed humans to survive the improbable chance of existence. From modern science, we currently understand that everything needed to work precisely together for life to exist on Earth. One example is the "gravitational constant" that, amazingly, matches the structures of our universe. Slightly more or less of this constant would make it implausible for galaxies and their planets to be formed as they currently exist.

Astrophysicist Ethan Siegel, who did his doctorate research on cosmological perturbation theory, wrote many scientific treatises for publications and web-sites[70] to introduce to the general public the latest research and discoveries about the universe.

Dr. Siegel remarked that, with respect to the matter and energy of the universe and the initial expansion rate, there are, so far, three possible options for the Big Bang events resulting in our universe.[71]:

1) The expansion rate could have been insufficiently large for the amount of matter-and-energy present within it, meaning that the universe would have expanded for a (likely brief) time, reached a maximum size, and then collapsed again. It's incorrect to say it would collapse into a black hole because space itself would collapse along with all the matter and energy, giving rise to a singularity known as the Big Crunch.

---

[70] https://www.forbes.com/sites/startswithabang/

[71] https://www.forbes.com/sites/startswithabang/2018/03/23/why-didnt-our-universe-collapse-into-a-black-hole, accessed June 24, 2019

2) The expansion rate could have been too large for the amount of matter and energy present within it. In this case, all the matter and energy would be driven apart at a rate too rapid for gravitation to bring all the components of the universe back together and, for most models, would cause the universe to expand too quickly to ever form galaxies, planets, stars, or even atoms or atomic nuclei! A universe where the expansion rate was too great for the amount of matter and energy contained within it would be a desolate, empty place indeed.

3) Finally, there's the "Goldilocks" case, or the case where the universe is right on the bubble between collapsing (which it would do if it had just one more proton) and expanding into oblivion (which it would do if it had one fewer proton). Instead, it just asymptotes to a state where the expansion rate drops to zero, but never quite turns around to collapse again.

It turns out the third option is the current and latest finding of what is happening to the universe as it exists today: the fine-tuned accuracy of the balance between the universe's expansion rate and its matter and energy density have resulted in our current place in the cosmos.

Looking at this new evidence in showing the sophisticated progression of the universe from the beginning to present time, I admire the Creator's divine wisdom in preparing this expanding universe in which our planet can exist. Even if a galaxy like the Milky Way was produced billions of years ago, the environment and living conditions of this home would not be ready for humans unless a delicate and thoughtful design was put in place.

An article[72] by Dr. Denny Lee tells us about our special privilege to live on a planet protected from harmful things like radiation. The Milky Way has a special "habitable zone" suitable for our solar system which, in turn, has another *habitable zone* specially tuned for the Earth so life can exist in its current state. Outside these zones, the environment would be too hazardous for life to start and survive. The Earth is miraculously shielded from harmful radiation and temperatures due to combinations of the right location and materials. This could never occur solely by random processes. Human beings are able to live on Earth because the Creator carefully designed and fine-tuned these conditions.

---

[72] Denny Lee, Double Protection for All Who Live on Earth, Challenger, Oct-Dec 2014. CCMUSA, http://ccmusa.org/read/read.aspx?id=chg20140402

Dr. Denny Lee told us about our special blessings to live on this planet Earth protected from harmful things like radiations. The discovery of Voyager 1 in 2012 by NASA showed that the radiation level of outer space, away from the inner protective atmosphere encircling the planet, was about hundred times more harmful. Such a high dose of radiation could wipe out all living things on Earth if it did not locate in the *habitable zone* and did not have the double safety nets of the protective atmosphere and its magnetic field. These scatter the majority of the cosmic radiation. The incredible biosphere we live in reflects the grace and wisdom of our Creator, God.

Also, Earth risks the constant bombardment of cosmic objects like asteroids, comets, and meteoroids. An astronomer, Professor David Bradstreet, shared the wonders and fascination of the uniqueness of our Earth:

> Our atmosphere deftly intercepts most space invaders. Every day our powerful atmospheric shield hijacks some 100 tons of small rocks and other pieces of space stuff heading our way, breaking up and incinerating everything before it can hit us. Even when larger cosmic visitors occasionally slip through our atmospheric defense system. God's sustaining processes of weather, wind and water erosion, volcanoes, and plate tectonics help Earth tidy everything up after we've been whacked. So successful are these planetary scrubbers that only in recent decades have scientists realized how often and how violently we've been struck in the past.[73]

Without the Earth's special shields from unwelcome objects, the cosmos's fly-by "bombs" would cause immense casualties to cities, forests, farm, rural areas, and our civilization.

Looking at the Moon and Mars, two of our neighbors, their surfaces are constantly bombarded by space objects that make craters all over their surfaces. A new study finds there are more than 200 asteroid impacts on the Red Planet, Mars, every year. These asteroids and comet fragments are usually no bigger than three to six feet (one to two meters) across.[74] Small space rocks burn up in Earth's atmosphere, never making it to the ground,

---

[73] David Bradstreet, Steve Rabey, Star struck: seeing the creator in the wonders of our cosmos, Grand Rapids, Michigan: Zondervan, chapter 15

[74] Pow! Mars Hit By Space Rocks 200 Times a Year, https://www.space.com/21198-mars-asteroid-strikes-common.html, accessed June 25, 2019

but they do damage on Mars because the planet has a much thinner atmosphere. The scientists with NASA's Near-Earth Object Program tracks some 600,000 orbiting objects, paying careful attention and analysis to 1,500 or so objects large enough and close enough to generate real damage.[75]

Information from NASA showed that:

> Every day, Earth is bombarded with more than 100 tons of dust and sand-sized particles. About once a year, an automobile-sized asteroid hits Earth's atmosphere, creates an impressive fireball, and burns up before reaching the surface. Every 2,000 years or so, a meteoroid the size of a football field hits Earth and causes significant damage to the area. Only once every few million years, an object large enough to threaten Earth's civilization comes along. Impact craters on Earth, the moon and other planetary bodies are evidence of these occurrences. Space rocks smaller than about 25 meters (about 82 feet) will most likely burn up as they enter the Earth's atmosphere and cause little or no damage. If a rocky meteoroid larger than 25 meters but smaller than one kilometer (a little more than 1/2 mile) were to hit Earth, it would likely cause local damage to the impact area.[76]

There are still unknown risks for what will happen in our refuge, Earth. The gracious protection from heaven to Earth, reminds me of a picture of a mother hen protecting her chicks[77] from the attacks of adversaries.

Also, as recorded in the Bible. "Surely he will save you from the fowler's snare and from the deadly pestilence. He will cover you with his feathers, and under his wings you will find refuge; his faithfulness will be your shield and rampart. You will not fear the terror of night, nor the arrow that flies by day..."[78]

---

[75] David Bradstreet, Steve Rabey, Star struck: seeing the creator in the wonders of our cosmos, Grand Rapids, Michigan: Zondervan, 2016: 174

[76] https://www.nasa.gov/mission_pages/asteroids/overview/fastfacts.html

[77] Matt 23:37; Luke 13:34

[78] Psalms 91:3 -5

The Bible[79] and current cosmological theory seem to agree on this: our world as we know it is headed for a rough future. Sometime in the future, the expansion of the universe will stop, and that could greatly impact the solar system and Earth. That may fill some people with anxiety, but professor and astronomer David Bradstreet is not worried.[80] "I've spent a lifetime examining the cosmos. I've seen the fingerprints of God in the wonders of Creation. I've witnessed his sustaining power holding everything together and making it all work. Why should I worry about how He's going to bring down the closing curtain? I'm sure He will do just fine."

Professor Bradstreet is looking for His Savior and Creator to bring him to a new heaven and earth[81]. "Our cosmos has been a pretty cool home, and every day we're learning more about exactly how cool it is. But a sequel is coming, and something tells me the new world will be even cooler than the old one."[82]

Just as essential as our Earth's special protection from a hazardous environment is our human body's defense system against harmful intruders. Besides the bacteria and viruses all over in our living environment on Earth, there are also various poisonous materials (both organic and inorganic) that can damage the human body internally and externally. The immune system and white blood cells inside the human body give us the first defense against microscopic intruders. Nevertheless, the most unique aspect distinguishing humans from animals is the special wisdom we possess for thinking, analyzing, organizing, and communicating. These unique talents help humans develop civilization and communities far more supreme than animals'. The knowledge of human beings in terms of physiological and medical development over thousands of years have given rise to medicines and technologies preventing sickness and treating disease. Within the last one hundred years, understanding about the science in DNA, cells, and organ functions have led to more healing to the physical body than previously thought possible.

Why is the human being so unique in the universe and Earth? Another astronomer and author of the book *Improbable Planet*, Dr. Hugh Ross,

---

[79] Revelation 6: 14

[80] David Bradstreet, Steve Rabey, Star struck: seeing the creator in the wonders of our cosmos, Grand Rapids, Michigan: Zondervan, 2016: 276–277

[81] Isaiah 65:17; Rev 21:1

[82] David Bradstreet, Steve Rabey, Star struck: seeing the creator in the wonders of our cosmos, Grand Rapids, Michigan: Zondervan, 2016: 277

showed with scientific evidence that human existence requires the harmonious coordination of astronomical, geological, chemical, and biological structures not made possible by mere accident.

Have you wondered why we are here and what is our purpose for being here? Dr. Ross shared his insight, "Humans tend to take pride and joy in all that we have achieved, and we have accomplished a great deal since the first of our kind set foot on Earth. Although we're inclined to take full credit, we must acknowledge that whatever we've attained sprang from the preparation and provision that proceeded us—on a cosmic, galactic, solar system, planetary, societal, and personal scale. Whatever we've done or become was made possible by a generous endowment of physical, intellectual, relational, and volitional capacities."[83]

In response to the wonderous habitability of Earth, people continue to ask more questions about the special and miraculous existence of human beings in the universe, such as: Are there aliens in outer space that may be similar or more intelligent than human beings?

I use scientific arguments to discuss this question and extend it to what might cause humans to exist in today's space and time.

1) Do aliens exist?

There are only two absolute answers to this question: "yes" or "no." If humans cannot answer firmly, they do not have enough evidence or confirmation to ensure a correct answer. The best they can respond at the moment is, "I do not know." Once we have enough scientific evidence—perhaps the detection of strong radio signals with meaningful communication from the outer space—the existence of aliens can be demonstrated. Such proof would show the existence of human beings in this universe is not unique.

An even stronger piece of evidence for alien existence would be objects or aliens arriving on our Earth that show us their civilization and existence. This, in reverse, is what we did in sending Voyager I and II to outer space in 1977. We expect other life forms might retrieve them and understand the messages. At that time, people on Earth relied on universally understood concepts, and developed a communication method neither specific to a location, nor a species, nor a world. Since the spacecrafts will orbit the Milky Way for the foreseeable future, NASA scientists decided that Voyager should carry a message from its "maker." They produced the *Golden Record*,

---

[83] Hugh Ross, Improbable planet: how earth became humanity's home, Grand Rapids, Michigan: Baker Books, 2016: 225

a collection of sounds and images they hope will be interpreted by other life forms. It contains[84]:

- 115 images of scientific knowledge, human anatomy, human endeavors, and the terrestrial environment.
- Spoken greetings in more than fifty languages.
- A compilation of sounds from Earth.
- Nearly ninety minutes of music from around the world.
- An English-language letter from former President Carter of the United States.

If the aliens in outer space sent out similar objects or an unidentified flying object (UFO), we could retrieve them and interpret communication from the intellectual beings.

Such evidence would verify the question, "Do aliens exist?" with an absolute "yes." However, we are biased toward a "no" answer because we have not found real evidence. To prove something is not in existence is generally more difficult because humans are limited in time and space. There is no way a human can search everywhere to make such a scientific claim that aliens do not exist. Only the Creator, who is above time and space, would know this for certain.

Using similar scientific arguments and principles, a second question can be asked of people with open minds who would like to investigate the facts and truth before answering another question.

2) Does God exist?

I used to be an atheist and believed in evolutionary theory until I spent time investigating the other side of the answer. Based on arguments similar to those in question 1, to prove God does not exist is much more difficult to prove than God exists. Some scientific theories and projections, based on basic experiments and fossils, seem to demonstrate that humans evolved from a lower order of mammals like monkeys, apes, or chimpanzees. However, the evidence is still superficial and not strong enough to provide a definite answer for the nonexistence of God. To say that "God does not exist" holds a scientific bias, just like saying "aliens do not exist."

A better approach is to investigate on both sides of the scientific argument before committing to a final answer and decisions. Analyzing how the

---

[84] https://voyager.jpl.nasa.gov/galleries/images-on-the-golden-record, accessed June 28, 2019

two answers are different, I see the "no" approach as a bottom-up investigation, while the "yes" approach is a top-down investigation as in Figure 8.

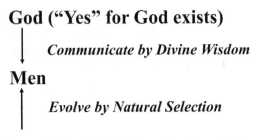

Figure 8: Diagram for "Does God exist?"

First, let's take the assumption of a top-down approach in that God created humans: How do we know He created us? A logical explanation would be a communication channel transmitted through the history of mankind, which can be received and interpreted by anyone who wants to listen. Since divine wisdom is far above Man's wisdom and our intellectual capability to comprehend, the Creator would find the best way to communicate to us directly or indirectly.

Watchman Nee in his book, *The Normal Christian Faith*, pointed out if God wishes to reveal himself to us, he must do it through human means. Human communication is conducted through speaking or writing, whether by telephone, signs, or symbols. Speaking or writing are the only two means by which God uses human language to directly communicate to humans. If he does so through writing, there must be one book which is divinely inspired among all the volumes written throughout history. If a divinely written book exists it must state explicitly that the author is God, carry a high tone of morality, describe the past and future of the universe in detail, and finally it must be available to humanity. These are the crucial tests if such book were to exist because it proves not only God's existence, but also his written revelation to us.[85] By checking against these qualifications

---

[85] Watchman Nee, *The Normal Christian Faith*, translated from the Chinese version book published by Church in Hong Kong Bookroom Ltd. Company, 1984: 28–37

to see if any book meets the four requirements, Watchman Nee[86] narrowed the choices down to the best book, the Bible.

It is interesting to note that these logical arguments of communication are like what the makers of Voyager I and II sent out in the *Golden Record* to reveal or introduce humans to aliens. In the case of the divine communication to humans, the maker is the Creator and God of the universe. God does not need to use a spaceship to transport the *Golden Record*, but He reveals Himself through the "Golden Book," the Bible, and the "living Bible," Jesus Christ.

On the other hand, using the assumption of a bottom-up approach for humans evolved from basic material: How do we know materials evolved into men? Philip Giorgio collected many good scientific studies about the beginning of the universe (13.8 billion years), bacteria (4 billion years), algae (1.2 billion years), mammals (0.3 billion years), and modern man (40K to 50K years ago).[87] The latest scientific argument for the evolution from material to human is by natural selection. Nevertheless, the farther back in time, the more challenging and difficult the investigation becomes. Just like a detective trying to find certain evidence for a crime scene, the farther away from the time of the crime—like one year, ten years, or one hundred years ago—the more difficult it becomes to collect valuable evidence.

I narrowed the figure in terms of time scale and opened the case for discussion and examination.

I developed a time scale using: A, B, C, D, Evolution theory, F, G, H, I… where A, B, C… refers to different theories about how humans evolved from material and mammals, etc.

With respect to a closer time scale, I picked the monkey (<300 million years ago) as a more appropriate representation in the above sequence, for the sake of checking the evidence.

A, B, C, D, Monkey, F, G, H, I…

Now, back to the top-down approach with similar arguments for a shorter time scale, I picked Jesus (~ 2,000 years) as the more appropriate representation in the above sequence for the sake of checking the evidence.

---

[86] Ibid, 37–48, content is taken from Watchman Nee 's Gospel message delivered in Tienstsin, China, in 1936. Nee kept his faith till his death in prison, witnessing Jesus Christ is the Son of God amidst the persecution of his faith.

[87] Philip Giorgio, Creation and the Arrow of Time, Xulon Press, November 11, 2017, 306–307

1, 2, 3, 4, Jesus, 6, 7, 8, 9, 10… where, 1, 2, 3… refers to different historical figures who claim they are the God whom humans came from. Figure 9 is a condensed version with a shortened time scale to zoom in and focus an investigation on the relevant evidence to determine the answer to the question, "Does God exist?"

Jesus (~2,000 years ago)

Communicate by Divine Wisdom

Men

Evolve by Natural Selection

Monkey (<300 million years ago)

Figure 9: Diagram for Jesus vs Monkey

If there is strong scientific evidence showing there is no divine intervention that caused a monkey or other life form to exist, including humans, the demonstration of showing no God in the universe is one step closer. Nevertheless, to prove the nonexistence of God is in general more of a challenge than to prove the existence of God. There exists some evidence, based on fossils and bones of primitive beings, that appears to trace the species between monkey and human beings by evolution. The latest DNA research among people of various ethical backgrounds can be traced it back to a single source of human-like species. Whether there was mutation from monkeys to evolve into that species is still not clear in terms of scientific evidence. In addition, there is a missing link to the existence of such a being somewhere between monkeys and human. Based on the latest investigations, the living interim species still cannot be found. I still have an open mind about this research and different hypotheses until more evidence is available.

On the other hand, people come up with different hypotheses that God created the evolutionary process itself in the beginning, and used it to shape what we have today. Nevertheless, wise person should start with dealing with the question of "Who" exists instead of "How" that relates to the so called "nothing" or "something" that exists during the beginning of the universe. I am a strong proponent of using science and evidence-based arguments to understand the "How" questions. Actually, the Biblical God does encourage people to use their mind to appreciate and investigate everything He created. The bottom line is that science is good for people when used correctly instead of misusing science to affirm the nonexistence of its Creator.

Looking at the evidence that Jesus left to the world, I found it much stronger, not only because of the shorter time scale (2,000 years to today), but also due to the number of people who witnessed the existence of Jesus. Had I not been introduced to the Bible where I met Jesus, I still would be an atheist or agnostic not wanting to have anything to do with God.

Regarding the original source of humans came from monkey through evolution or God's creation through Jesus, I leave it to your fair choice after reviewing the evidence. Besides the historical evidence, there is creation evidence and Biblical evidence allowing you to explore the God of wonders[88] in nature and the God of love in the supernatural. Just like the beginning of the universe was supernatural, the death of Jesus on the cross was a demonstration of God's nature of love to restore the communication channel supernaturally. In essence, God's nature is supernatural to human understanding. Most important of all, the evidence can help you communicate with Jesus, according to His Word, and to check out if He is real or not. The next chapter shares more about the love in God's nature, which we can apprehend indirectly through Biblical revelation and directly through Jesus Christ.

---

[88] DVD presentation, God of Wonders: Exploring the wonders of creation, conscience, and in the glory of God, Eternal Productions, 2010, one of the examples in the topic for God of Wonders

THE WISDOM COMMUNICATED
THROUGH DIVINE NATURE

Chapter Ten

LOVE AND LIGHT

T he mighty works and glory of the God of the universe are witnessed through the amazing structures of heaven and Earth. The more scientists discover about how marvelous the universe is, the more people see the wonders of the creation. The general revelation of nature shows us the power of the Creator. Nevertheless, people may ask, "What is God?" or "How do I know if God exists so I can communicate with Him?" Different religions tried to tell people who is God and what is God throughout the civilization of mankind. Specific revelations from the Bible communicate to us about God's nature and who He is. Just like we engraved some pictures and songs in the "Golden Record" of Voyager I and II, spaceships sent to the outer space to introduce who humans are, God tells us who "God is" through the Bible.

By searching the Bible with the exact key words of "God is," I can learn directly what the divine nature is. There are certainly more places in the Bible telling us His nature: stories, Psalms, Proverbs, letters, etc. The direct declarations of the divine nature are the best introductions to God so we can know Him better, and the relationship between God and humankind can be built. Even on a human level, you need to know a little more about a person's character, nature, likes, dislikes, etc. before you can establish relationships. These include friendships, teachers/students, boyfriend/girlfriend, husband/wife, etc.

I did not know my father very well when I was young because he worked about 360 days a year, including most weekends. The Chinese culture at that time also kept most fathers from expressing themselves in terms of his loving relationships. Their way to show love to their children was to provide enough food, education, and a living for the family. Not until I became a father, did I understand a little more about his nature by spending some time with him when I visited Hong Kong. However, I still missed the "golden opportunity" to know my father when I was young because he was always busy and not available.

I learned the lesson of not repeating the same mistakes. I made sure to spend reasonable and enough time with my children when they were under twelve years old. After that age, their tendency and preference turn to spending more time with their friends. But time shared with them while young will last as memorable moments for life. In that sense, time is life and life is the meaningful moments people spend together. A Chinese proverb says, "An inch of time for an inch of gold, yet an inch of gold cannot buy an inch of time." The wisdom is that the value of the immaterial essence of time, exceeds that of the most precious substance in the world.

On the other hand, God is always available and He is waiting for us to spend time with Him in communicating, fellowship, walking, serving, and building relations. To spend time effectively in knowing God, we can build a good relationship with Him by studying the Bible regularly. With a computer readily at hand, people nowadays can learn the divine nature systematically and efficiently.

I chose seven verses in the Old Testament and seven verses in the New Testament of the Bible to represent the introduction of the divine nature. I encourage more searching in other verses to help understand God's nature even more in-depth.

Seven verses in the Old Testament:

1. God is holy   Psalm 99: 9

2. God is full of compassion   Psalm 116: 5

3. God is the Lord   Psalm 33: 12

4. God is our God for ever and ever   Psalm 48: 14

5. God is merciful and forgiving   Daniel 9: 9

6. God is righteous in everything He does   Daniel 9: 14

7. God is gracious and compassionate   2 Chronicle 30: 9

<u>Seven verses in the New Testament:</u>

1.	God is truthful	John 3: 33
2.	God is spirit	John 4: 24
3.	God is faithful	1 Cor. 1: 9
4.	God is alive and active (His Word)	Hebrews 4:12
5.	God is light	I John 1:5
6.	God is love	I John 4: 8
7.	God is One	Mark 12: 32

The love of God is magnified by His divine nature in loving and caring for His children. Just like parents would prepare and take heartfelt steps to establish a home in which their children can grow up, God created the heaven and Earth in proper sequence so humans can live in the proper environment.

The greatest love from God, that surpasses all the external things He prepares and gives to us, is communicated to us in John 3:16[89]: "For God so loved the world that he gave his one and only Son, that whoever believes in him shall not perish but have eternal life." He died for our sins so we can live. This is the greatest Love in the entire universe.

The Bible tells us humans were created in the image of God, we bore His likeness in character and nature. One of the greatest human gifts is our free will to choose, along with the creativity, passion, compassion, artistry, an analytical mind, and so on. The history and stories recorded in the Bible nicely illustrate God's relationship with mankind.

Adam and Eve were placed in the Garden of Eden. They and their off-spring were supposed to "be fruitful and increase in number; fill the earth and subdue it. Rule over the fish in the sea and the birds in the sky and over every living creature that moves on the ground."[90] The communica-

---

[89] The processes of God's creation and redemption are shown throughout the entire Bible, where John 3:16 is the highlight and summary of the main theme of the Bible.

[90] Genesis 1:28b

tion between God and humans was good. Adam and Eve understood what God told them to do and not to do. The tree of knowledge of good and evil was the only condition that would result in death and separation from God. Because of free will and temptation, Adam and Eve did not obey God's Word and rebelled against it.

Taking the fruit of the tree of knowledge of good and evil, Adam and Eve not only excommunicated themselves from God, but also passed on their sin to the entire human race. There was no way out of the law of sin or a way to restore relations with God through human effort and endeavor. The poison of sin not only damaged the physical bodies, but also the souls of human beings. The spirit of mankind was basically dead, as God told Adam and Eve what would happen if they took the fruit from the forbidden tree.

Jesus Christ, the Son of God, is the only remedy and spiritual prescription to recover the relationship that God originally created with mankind. "Therefore, there is now no condemnation for those who are in Christ Jesus, because through Christ Jesus the law of the Spirit who gives life has set you free from the law of sin and death. For what the law was powerless to do because it was weakened by the flesh, God did by sending his own Son in the likeness of sinful flesh to be a sin offering."[91]

Just like the law of aerodynamics can overcome the law of gravity in moving people upward, the law of the Spirit can overcome the law of sin and death in saving mankind. The human relationship with the law of Spirit and Jesus Christ, opens the gate for people to enter God's kingdom.

The messages of peace, love, and hope have been spreading to all human beings through the Gospel of Jesus Christ. Apart from Christ, the devil and sin dominated all of the human race. The contrast between the two "kings"— Christ for the kingdom of God, and devils for the kingdom of Satan—is light and darkness respectively. Without the true light of Christ, mankind constantly lived in wars, fear, bondage, and lies to one another. The darkness in the world does not like to receive the true light. History shows sin has influenced all cultures including the East, the West, Africa, and the Middle East. The Bible also recorded how Israel repeatedly failed to obey God's law and even worshipped the idols of Canaan. Israel, as God's chosen people, were a model to represent His salvation plan. When they did not follow God's Word, trouble and deterioration of morals and their kingdom resulted. Eventually, the country of Israel was destroyed more than 2,500 years ago, until last century.

---

[91] Romans 8:1–3

Only when the Word of God lives in the heart of men, can men truly overcome sin. The living Word of God is Jesus Christ. He is the Messiah not only for Israel's people, but also the Savior to all humankind.

People may ask, "Why did God allow the tree of knowledge of good and evil to be in the garden of Eden, or allow Adam and Eve to choose the fruit of forbidden tree?" Reviewing the answers to these fundamental questions of God's sovereign and mankind's free will, we understand better the love of God. Could God not put the tree of knowledge of good and evil in the garden so Adam and Eve would not have the choice to take the fruit? They could then happily live forever in the paradise and their children would enjoy all the goodness on Earth. Another option would be if God didn't create other races with different religious practices in Canaan so Israel was not able to worship other idols and gods? The Ten Commandments, given to the Israel through Moses, forbade them to worship other gods. Jesus wisely answered these questions using three parables to help people apprehend the essence of the love of God. One of them is the famous parable of the "prodigal son."

> There was a man who had two sons. The younger one said to his father, "Father, give me my share of the estate." So, he divided his property between them. Not long after that, the younger son got together all he had, set off for a distant country and there squandered his wealth in wild living. After he had spent everything, there was a severe famine in that whole country, and he began to be in need. So he went and hired himself out to a citizen of that country, who sent him to his fields to feed pigs. He longed to fill his stomach with the pods that the pigs were eating, but no one gave him anything. When he came to his senses, he said, "How many of my father's hired servants have food to spare, and here I am starving to death! I will set out and go back to my father and say to him: Father, I have sinned against heaven and against you. I am no longer worthy to be called your son; make me like one of your hired servants." So he got up and went to his father. But while he was still a long way off, his father saw him and was filled with compassion for him; he ran to his son, threw his arms around him and kissed him. The son said to him, "Father, I have sinned against heaven and against you. I am no longer worthy to be called your

son.' But the father said to his servants, 'Quick! Bring the best robe and put it on him. Put a ring on his finger and sandals on his feet. Bring the fattened calf and kill it. Let's have a feast and celebrate. For this son of mine was dead and is alive again; he was lost and is found."[92]

The father should have had the authority to not give his younger son his share of the estate, because the father was still alive and could decide what his sons got from him. The father could have initially chosen to lock his younger son in his home, against his will to leave home. But the father made a surprising decision, giving his son the fortune. He let him go because the son could not resist the temptation of the outside world and chose to leave home. The son probably thought he would enjoy life better outside his home without the love of his father. The father allowed his son the free will to choose, even though the father knew what would happen. If not allowed to leave home, his son would not be happy and would think of the outside world as always better than home.

To capture his son's heart, the father let his younger son have the freedom. He himself, though, would suffer in the absence of his son. The material and money loss were secondary to the father, but the loss of his son caused major distress because the father so loved his son. This is reflected in the description of how the father waited for his son to come back home, and his excitement when he ran to greet his son compassionately. The father even embraces his son as family member again, instead of rebuking him or treating him as servant.

This is exactly how much love our Heavenly Father holds for mankind. He waits for His beloved children to repent and turn back to God to restore the relationship broken previously. The consequence of the prodigal son in leaving home through his free will is an analogy for the consequence of our ancestors in choosing the fruit of the tree of knowledge of good and evil through their free will. However, it is also the salvation for those who choose to open their hearts to come back to their heavenly Father's home.

The picture of the triune God is also communicated through the three parables of Luke 15. The parable of the shepherd looking for the lost sheep is an analogy of God the Son coming to the world to look for the lost people (sheep). The parable of the woman searching for a lost coin is analogized to

---

[92] Luke 15:11–32

God the Spirit shining spiritual light into the darkness of the human heart to enlighten the lost person. The parable of the father waiting for the lost son is an analogy of God the Father forgiving the lost person because he or she repented. These three parables, in one main theme, demonstrate the Love of God in the action of reaching out to all human beings. Truly "God is Love" is more than a theory or religious practice. The wisdom and goodness of God's Love is fully communicated through our Lord Jesus Christ.

How will you respond to this amazing Love and Grace?

People may more easily understand "God is Love" than "God is light," per 1 John 1:5. But they may also ask, "What does it mean that God is light?"[93] How does that relate to me when God is light? How can I see and understand that "God is light"? Does this light refer to a physical or spiritual aspect?

Light is a common metaphor in the Bible. Proverbs 4:18 symbolizes righteousness as the *morning sun*. Philippians 2:15 likens God's children, who are *blameless and pure*, to shining stars in the sky. Jesus used light as a picture of good works: "Let your light shine before others, that they may see your good deeds" (Matthew 5:16). Psalm 76:4 says of God, "You are radiant with light."

The fact that God is light sets up a natural contrast with darkness. If light is a metaphor for righteousness and goodness, then darkness signifies evil and sin. First John 1:6 says, "If we claim to have fellowship with him and yet walk in the darkness, we lie and do not live out the truth." Verse 5 says, "God is light; in him there is no darkness at all." Note that we are not told that God is *a* light but that He *is* light. Light is part of His essence, as is love (1 John 4:8). The message is that God is completely, unreservedly, absolutely holy, with no admixture of sin, no taint of iniquity, and no hint of injustice. If we do not have the light, we do not know God. Those who know God, who walk with Him, are of the light, and walk in the light. They are made partakers of God's divine nature "having escaped the corruption in the world caused by evil desires" (2 Peter 1:4).

God is light, and so is His Son. Jesus said, "I am the light of the world. Whoever follows me will never walk in darkness, but will have the light of life" (John 8:12). To *walk* is to make progress. Therefore, we can infer from this verse that Christians are meant to grow in holiness and to mature in faith as they follow Jesus.[94]

---

[93] https://www.gotquestions.org/God-is-light.html

[94] Ibid

The above answers to the question about "God is Light" match very well to the first paragraph of John 1: "In the beginning was the Word, and the Word was with God, and the Word was God. He was with God in the beginning. Through him all things were made; without him nothing was made that has been made. In him was life, and that life was the light of all mankind. The light shines in the darkness, and the darkness has not overcome it."[95]

This same Word is the Word spoken by God to create the universe and, specifically, the light in the first day of creation in Genesis. "And God said, "Let there be light," and there was light."[96] Nevertheless, the most amazing thing that happened in this universe is the same Creator and Great God came to this small Earth as a human being through incarnation! He started His human life as a small baby born in Bethlehem about 2,000 years ago. Just like John 1:14 declares: "The Word became flesh and made his dwelling among us. We have seen his glory, the glory of the one and only Son, who came from the Father, full of grace and truth."

The beginning of His human life is unusual in that He was conceived through the Spirit unto a virgin woman name Mary. "This is how the birth of Jesus the Messiah[Savior] came about: His mother Mary was pledged to be married to Joseph, but before they came together, she was found to be pregnant through the Holy Spirit."[97]

Mary's response to the angel's announcement also shocked her. "You will conceive and give birth to a son, and you are to call him Jesus. He will be great and will be called the Son of the Most High. ... 'How will this be?' Mary asked the angel, 'since I am a virgin?' The angel answered, 'The Holy Spirit will come on you, and the power of the Most High will overshadow you. So the holy one to be born will be called the Son of God....'"[98]

It was different from our current knowledge of how human beings are conceived through the flesh physically[99]. This special pregnancy seems as mysterious as how the universe started from the beginning. Science is still not able to explain with convincing theories how that could have happened. How can physical material come into existence from nothing if there was a

---

[95] John 1:1–5

[96] Genesis 1:3

[97] Matt 1:18

[98] Luke 1:31–38

[99] Fertilization is the joining of a sperm and an egg; https://en.wikibooks.org/wiki/Human_Physiology/Pregnancy_and_birth, accessed July 16, 2019

beginning? Before the beginning of the physical universe, there was nothing based on the latest science modeling, including the "Big Bang" theory. However, the Bible told us that God already existed before the beginning of the universe. God is Spirit, who is in the spiritual dimension and above the physical dimensions. Only a spiritual being of God, who exists from eternal past to eternal future and beyond the limitation of time dimension, can have the mighty power and wisdom to bring forth something physically from nothing. This is a logical argument that all existing physical things, including life, have to come from something physical. For instance, in the natural realm, men and women come from their father and mother, who also come from their parents, and so on. If there is nothing physical at the beginning, nothing physical can follow based on human reasoning. Only if there was Spirit—above time and space, and not limited by physical things—could the heaven and Earth start and exist as we see it today. The Bible clearly reveals this amazing truth to us about God's creation.

In addition to needing the important spiritual dimension for light that brings forth spiritual life in Christ's followers, physical light is essential for all physical life to grow on Earth. From plant and animal life to human life, physical light is needed to make food and vitamins so all life can survive within the food chain and cycles. That could be the reason light first appeared in Genesis' creation account through God's Word. "And God said, 'Let there be light,' and there was light. God saw that the light was good, and he separated the light from the darkness. God called the light 'day,' and the darkness he called 'night.' And there was evening, and there was morning—the first day."[100]

Physical light by itself also possesses two amazing natures: particle and wave. Through the work of Max Planck, Albert Einstein, Louis de Broglie, Arthur Compton, Niels Bohr, and many others, current scientific theory holds that all particles exhibit a wave nature, and vice versa.[101] This phenomenon has been verified not only for elementary particles, but also for compound particles like atoms, and even molecules. Macroscopic particles' wave properties, because of their extremely short wavelengths, usually cannot be detected. We can calculate the wave equation to determine its frequency or wavelength based on the velocity of the wave: $V = f * l$

---

[100] Genesis 1:3–5

[101] Walter Greiner (2001). Quantum Mechanics: An Introduction. Springer. ISBN 978-3-540-67458-0

Where V is the velocity of the wave in m/s; f is the frequency in Hz and l is the wavelength of the wave in m. Waves can have constructive interference or destructive interference, which are characteristic of waves. This phenomenon can form standing waves, harmonic noise cancelation, etc., using different phase conditions of the wave.

In 1905, Albert Einstein provided an explanation of the photoelectric effect that shows light has the nature of a particle. He explained that the electrons can receive energy from an electromagnetic field (including light) only in discrete units (quanta or photons): an amount of energy E that was related to the frequency f of the light by $E = h*f$ where h is Planck's constant (6.626×10^{-34} Js).

Only photons of a high enough frequency—above a certain threshold value—could knock an electron free. For example, photons of blue light have sufficient energy to free an electron from metal, but photons of red light do not. One photon of light above the threshold frequency could release only one electron. The higher the frequency of a photon, the higher the kinetic energy of the emitted electron, but no amount of light below the threshold frequency could release an electron. This represents the quintessential example of wave-particle duality. Electromagnetic radiation propagates following linear wave equations, but can only be emitted or absorbed as discrete elements, thus acting as a wave and a particle simultaneously.[102]

The dual nature of physical light revolutionized the thinking and perception of science in the last century, when the classical concept only allowed one or the other. Nowadays, scientists will no longer consider wave as wave only and particle as particle only. Wave-particle duality is exploited in electron microscopy, where the small wavelengths associated with the electron can be used to view objects much smaller than we can see in visible light. Photos are now able to show this dual nature, which may lead to new ways of examining and recording this behavior.[103]

The beauty in the characteristic of light reflects the glory of God through Jesus Christ. He is even far more supreme than physical light because He is the light of life (John 8:12) in the spiritual dimension. He has the dual nature of both God and man. This revelation seems even more difficult to understand than the physical nature of the dual property of light as both wave and particle. If the new concept of light blew the mind of the scientists

---

[102] https://en.wikipedia.org/wiki/Wave-particle_duality, accessed on July 17, 2019

[103] https://phys.org/news/2015-03-particle.html, accessed July 17, 2019

last century, people were even more amazed to learn that Jesus Christ was both God and man.

We need wisdom to understand one of the most important natures related to God and Jesus Christ. As mentioned by a MIT professor and a plasma physicist, Ian Hutchinson, "Elaborate analogy and metaphor are essential to convey the theological significance of our relationship to Christ."[104]

For example, on a human level, it is correct to say that I have both my father and mother's natures. The dual nature can be seen not only partly in my external appearance but also my internal character. I do carry the genes of both of my parents. That I like to think of and evaluate new ideas and concepts resembles my mother's side. I also have the appearance of and I function like my father's side. Of course, I developed my own character and distinct passions to do things as I grew up. The foundation, however, started from the dual nature of my father and mother when I was born.

Back to the divine level, because Jesus Christ has the dual nature of God and men, He can open the channel of communication between the divine wisdom and human beings. That is why the key to building a valid relationship with God is through God's Word. Jesus Christ is the Living Word of God as revealed from the Bible. Later, we will further discuss the communication of Jesus Christ to human beings using the T transform we covered before. This will give you a map to help find the treasure of meeting Jesus Christ, as told in Bible verses. If you can communicate back to Jesus Christ, your spiritual eye will be opened to see how real and amazing the Creator and Redeemer is.

"In the past God spoke to our ancestors through the prophets at many times and in various ways, but in these last days he has spoken to us by his Son [Jesus Christ], whom he appointed heir of all things, and through whom also he made the universe. The Son is the radiance of God's glory and the exact representation of his being, sustaining all things by his powerful word."[105]

---

[104] Ian Hutchinson, Can a scientist believe in miracles?: an MIT professor answers questions on God and science, Downers Grove, InterVarsity Press, 2018: 192

[105] Hebrews 1:1–3

Chapter Eleven

SPIRIT AND WISDOM

B efore Jesus's crucifixion, He told His disciples about the work of the Holy Spirit. "When the Advocate comes, whom I will send to you from the Father—the Spirit of truth who goes out from the Father—he will testify about me...."[106]

What does it mean that "the Spirit of truth" will testify about Jesus? Let's review some relevant information to better understand this idea.

"Spirit" is mentioned more than 500 times in the Bible, approximately 200 times in the Old Testament and more than 300 times in the New Testament. Genesis 1:1's first creation verse, "In the beginning God created the heavens and the earth," is immediately followed by a verse that describes the Spirit of God, and the third verse describes God's speaking. "Now the earth was formless and empty, darkness was over the surface of the deep, and the Spirit of God was hovering over the waters. And God said, "Let there be light," and there was light."[107]

It seems there is a relationship between the Spirit and God. A more straightforward revelation to this relationship is given by Jesus in John 4: 24, "God is Spirit, and those who worship Him must worship in spirit and truth." [108]

This statement—with uppercase Spirit equal to God's Spirit, and lowercase spirit referring to the humans'—is a revolutionary concept about God and how the true worshipper can worship Him. The implication of the above Biblical verses shows that God is not limited by physical space and time. Religious people usually think the worshippers need to go to certain mountains or temples to worship God. However, God is above physical

---

[106] John 15:26–16:15

[107] Genesis 1:2–3

[108] John 4:24, NKJ Bible

limitations. He was telling people the most important means for them to worship Him is in spirit and truth. In the Old Testament, God uses physical things to help people understand what the "spiritual things" are all about. Some of the material objects recorded in the Bible foreshadow "spiritual things" and "spiritual dimensions" in heaven. The reasons could be due to the fact that human beings get used to things on Earth and live a life bounded by space and time. For example, the sanctuary built by the Israelis and the temple built by King Solomon were models representing genuine things in heaven.

This is explained more specifically in Hebrews 8:5–6: "They [high priests] serve at a sanctuary that is a copy and shadow of what is in heaven. This is why Moses was warned when he was about to build the tabernacle: 'See to it that you make everything according to the pattern shown you on the mountain.' But, in fact, the ministry Jesus has received is as superior to theirs as the covenant of which he is mediator is superior to the old one, since the new covenant is established on better promises."

Even King Solomon's prayer of dedication for the temple he built express this. "But will God really dwell on earth with humans? The heavens, even the highest heavens, cannot contain you. How much less this temple I have built!"[109]

God is Spirit and is definitely greater than the physical realm of the heaven and the Earth. In Isaiah 57:15, the Bible further tells us that the Lord God lives in a place and in a time different from humans, but He choose to live with those who are repentant and humble in spirit. Human beings live in a realm of time and space that is so limited. They are not able to approach God without a humble spirit. This agrees with what Jesus told us: we can only worship God through the human spirit and truth instead of physical things.

When I was young, my older brother, Ming, and I liked to put together model boats with store-bought components. Ming liked to assemble the appearance or external decorations on the model boat, while I liked to assemble its internal motor.

Once the assemblies were finished, we brought the model boats to a local park with a small boat pond of about 50 m² (~ 440 ft²). The model boats appeared and moved like real boats, but could never truly replace one. The external decorations were full of plastic cannons, decks, soldiers, etc. People would need to use their imaginations to transform those

---

[109]2 Chronicles 6:18

nonfunctional items into the real functionality and usage of real boats. Someone who had never seen a real boat before, however, could get a glimpse of how one would look. Inside the model boat, a motor, wires, battery, and other items needed to be connected exactly, per the assembly instructions. If the battery polarity was reversed, the model boat would not move correctly. I liked to investigate different ways to improve the boat's power and movement, like reducing the water resistance, or waterproofing the model boat. No matter how I improved the design, I knew it would not function as a real boat. The actual boat, with a much bigger engine, operated in a more superior way. The small motorboat may have a similar outward appearance and the likeness of an engine found inside a real boat. However, the model boat was simply a shadow of reality.

In the same way, we can gain insight, and enhance our visualization of true spiritual things, by learning how corresponding earthly things look and operate.

Take the example of King Solomon's temple mentioned earlier. The magnificent structure was one of the greatest buildings some 3,000 years ago. The physical temple building was destroyed about 2,600 years ago. Nevertheless, the actual meaning and function of the temple pointed people to the real treasure of the spiritual temple, and that can never be destroyed.

"Consequently, you are no longer foreigners and strangers, but fellow citizens with God's people and also members of his household, built on the foundation of the apostles and prophets, with Christ Jesus himself as the chief cornerstone. In him the whole building is joined together and rises to become a holy temple in the Lord. And in him you, too, are being built together to become a dwelling in which God lives by his Spirit."[110] The genuine temple, where God dwells, is not assembled by physical things and earthly materials but through spiritual things. This spiritual temple was also described by the apostle Peter, "As you come to him, the living Stone— rejected by humans but chosen by God and precious to him—you also, like living stones, are being built into a spiritual house [or into a temple of the Spirit] to be a holy priesthood, offering spiritual sacrifices acceptable to God through Jesus Christ."[111]

The above Bible verses are just a few illustrations showing how to model the physical temple to reflect the spiritual and holy temple, where God would love to dwell. In the genuine temple, His people can be united

---

[110] Ephesians 2:19–22

[111] I Peter 2:4–5

with Him through the Spirit and Truth. They will serve and worship Him forever and ever.

In contrast with the spiritual house, where God's household is being built, a physical temple creates more of a physical separation keeping people from approaching God. People are not holy in the sight of God, who is Holy and righteous. No way can people come to God and be cleansed completely by a periodic physical offering. The outward Mosaic commandments and laws help prepare people to approach God if they keep them. If no human being can keep them, the law of God can condemn every person with a reminder of his or her sin. History has shown that the flesh of all mankind tempts them to rebel against God and His Word. The flesh is the opposite of the Spirit in terms of pleasing God and following His Word.

The Apostle Paul experienced this first-hand in his struggle with his weakness in keeping the outward law. Even though he knew God's law was good, he would not follow it. Instead he followed the evil desire of the flesh.[112] The famous doctrine in the book of Romans about the salvation through Christ clearly demonstrates the mighty power of the Spirit operated in the believers:

> Those who live according to the flesh have their minds set on what the flesh desires; but those who live in accordance with the Spirit have their minds set on what the Spirit desires. The mind governed by the flesh is death, but the mind governed by the Spirit is life and peace. The mind governed by the flesh is hostile to God; it does not submit to God's law, nor can it do so. Those who are in the realm of the flesh cannot please God. You, however, are not in the realm of the flesh but are in the realm of the Spirit, if indeed the Spirit of God lives in you.[113]

The flesh is aligned to the physical realm with time and space limitations, whereas the Spirit is aligned to the spiritual realm without time and space limitations. The divine wisdom is communicated to us through Jesus Christ, who was offered as an eternal sacrifice to take away the sins of the world. God's Love and Light can be demonstrated and expressed because of Christ's ministries on Earth from Jesus' birth to His death in the cross. Only

[112] Romans 7: 14–24

[113] Romans 8: 5–9a

Jesus's dual nature of divine and human allowed him to overcome death. He resurrected in life to send the Spirit of God to the people who belong to Him.

How can people belong to God? Are the Israelites the chosen people who belong to God? Again, the model of Israelites in the physical realm as God's people was expanded into the boundless spiritual realm for all mankind, who receive Jesus as his or her savior and Lord. The original plan of God was to establish a unique relationship for mankind to be truly united with God in the spiritual realm. The amazing blueprint of salvation, sanctification, and glorification were fully carried out by Jesus Christ.

During a conversation with Nicodemus, a member of the Jewish ruling council, Jesus revealed this blueprint effectively:

> Jesus replied, "Very truly I tell you, no one can see the kingdom of God unless they are born again."
>
> "How can someone be born when they are old?" Nicodemus asked. "Surely they cannot enter a second time into their mother's womb to be born!"
>
> Jesus answered, "Very truly I tell you, no one can enter the kingdom of God unless they are born of water and the Spirit. Flesh gives birth to flesh, but the Spirit gives birth to spirit. You should not be surprised at my saying, 'You must be born again.' The wind blows wherever it pleases. You hear its sound, but you cannot tell where it comes from or where it is going. So it is with everyone born of the Spirit."[114]

The human spirit is a unique part of all human beings and was originally placed inside them by God for the sake of direct communication with God. However, our ancestors listened to lies from the crafty serpent[115] that resulted in spiritual death and excommunication from God. The human spirit then became nonfunctioning and was replaced by the flesh, which gave rise to fleshly beliefs for all humans. This fall caused human flesh to became waste and void in the sight of God, and darkness fell upon the human heart. Therefore, the Bible says that humanity's nature, both the physical and spiritual, were good originally. The human nature was harmfully affected

---

[114] John 3:3–8

[115] Genesis 3:1–13

by sin after the fall. The outcome of sin is a fallen human nature, striving against God, and pursuing sinful indulgence, typically referred as the "flesh" in Scripture. This condition is similar to the description at the beginning of creation, if you compare the human heart to the Earth: "And the earth was waste and void; and darkness was upon the face of the deep: and the Spirit of God moved upon the face of the waters. And God said, let there be light: and there was light."[116]

The Earth followed exactly what God said: "Let there be light." Nevertheless, humans were created in the image of God, and every human being has the free will to choose to accept or reject the light of life. As soon as people turn their hearts to God and accept Jesus Christ into their hearts, the Spirit of God dwells within a human's spirit. The spiritual light from God will enlighten the spiritual eye of the believers to see the true light, Jesus Christ. They can be the children of God[117] and are truly spiritually reborn into the kingdom of God. God's Love and Light will shine into the human heart so he or she can walk in the love and light that expresses God's.

In summary, the Spirit of God and the Word of God give rise the "old creation" as descripted in Genesis 1. The Spirit of God is the essence and attributes of God, whereas the Word of God is the expression of God that became flesh as Jesus Christ. However, Jesus has the likeness of flesh but, with both a divine and a human nature, he was qualified as the redeemer to save people. He had the dual nature of God and man, but without sin[118]. He is the way, the truth and the life[119] for the "New Creation"—showing human beings can be born again spiritually and enter eternally into the kingdom of God. Jesus is the King of the kingdom of God and He rules in the kingdom so His people can dwell in the house of the Lord forever with goodness and love.

The famous shepherd's Psalm 23, written by King David about 3,000 years ago, gives us a good model and vision of what the kingdom of God will look like with the Lord as a good shepherd. This Great Shepherd, the Lord Jesus Christ, loves His sheep (people who belong to Him), makes them lie down in green pastures (paradise), and leads them beside quiet waters

---

[116] Genesis 1:2–3, ASV Bible

[117] John 1:9–12

[118] 1 Peter 2:22; Hebrews 4:15

[119] John 14:6

(life supply). The Lord refreshes their souls and guides them along the right paths because He is with them (Emanuel).

In the New Testament, the visions are even clearer for us to understand because the Messiah and Savior of the world, the Lord Jesus Christ had come. The life stories, ministries, teachings, and accomplishments of Jesus Christ—about 2,000 years ago—were recorded in that part of the Bible. The testimonies about Him were overwhelmingly documented in historical records by personal witnesses of that age, and historians throughout the ages, in addition to the Bible. Jesus's physical appearance on Earth in the past and his interaction with human beings was like a good shepherd to lead his sheep away from the valley of death.

Through His death on the cross, and his resurrection, Jesus Christ can now live in the hearts of all His followers through the Spirit. The spiritual witnesses related to Jesus Christ and His impact to the world was recorded in the Bible and in other testimonies throughout many generations and into current times. The Bible verses in 1 Timothy 3 tell us that, "God's household, which is the church of the living God, the pillar and foundation of the truth. Beyond all question, the mystery from which true godliness springs is great: He appeared in the flesh, was vindicated by the Spirit, was seen by angels, was preached among the nations, was believed on in the world, was taken up in glory."[120] This amazing truth about God appeared in the flesh, Jesus Christ, not only revolutionized the outward religions, but most importantly, transformed human hearts from sinful conditions of devils into the holy conditions of God.

Figure 10 is a roadmap to connect key terms and show their relationships. The arrow can be deemed as an expression from one term to the other term based on the direction of the arrow. Without the revelation and understanding of the key relations, we could hardly know God. We would think He was too abstract or too far away for us to know. We might even use human perceptions to project what God would look like according to what we think.

Per Biblical investigation, we can avoid the error of a wrongful description of God. Please use this map to help you discover for yourself even more precise and accurate information about God. Often, a map will help you track down the destination more effectively, or make your path easier to find life's hidden treasure. Alternatively, you can go straight to the Bible's four Gospels to learn more about God.

---

[120] 1 Timothy 3:15b–16

Starting from God at the top of the picture in Figure 10, the terms below God are often described in the Bible. It says "God is Spirit" and "the Word is God," so I used two arrows for Spirit and Word to express God. The Word (Jesus Christ) became flesh and Christ was conceived by the Holy Spirit. Note that Jesus is an English name, meaning "God is salvation," the same as Joshua in the Hebrew language. Christ is a Greek word meaning "the anointed one," or "the chosen one," corresponding to the title or role of Jesus. Christ means "Messiah" in the Hebrew language, and "Savior" in common English terms.

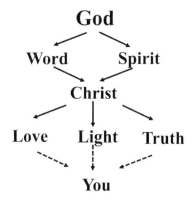

Figure 10: Key word highlight map

Christ came to fully reflect and express the divine character of God, who cannot be seen and approached directly by human beings. The Bible told us "God is Love, "God is Light," and "God's Word is Truth." All these divine characteristics are fully demonstrated, illustrated, and magnified through Jesus Christ, who has the dual nature of God and man. Jesus completely carried out His mission on Earth with passion and love. His teaching started with telling people to repent as the kingdom of God was near[121]. Through suffering on the cross, death, and resurrection, Jesus accomplished His mission for spiritual redemption, which opened the door for believers to enter the kingdom of God. Consequently, the Spirit of God can communicate directly to the spirit of all Christ followers. This gospel (good news) has been shared with all people and all nations, so people can respond to and believe in the gospel. These messages reveal the real wisdom from God.

---

[121] Mark 1:15

How do you respond to these important and beautiful messages communicated to you from God? You can gain wisdom by knowing and understanding the messages from God.

PART FIVE

THE WISDOM COMMUNICATED THROUGH SPECIAL REVELATION

Chapter Twelve

THE BIBLE

The Bible starts with God's creation and God speaking in Genesis 1:3: "And God said, "Let there be light," and there was light." After He spoke, miracles happened. Starting with light, water, space, and life, the law of physics governed all of creation. God had a purpose in speaking; He wanted to communicate to those who understood about Himself, just as an artist creates a painting or sculpture working in thoughts and messages to communicate to viewers. He or she will be happy if someone understands the message behind the art. Similarly, God not only is pleased when His people know Him and His will, He also wants to establish an eternal relationship with them.

Fast forward to the end of the Bible to Revelation 22:17. "The Spirit and the bride say, 'Come! And let the one who hears say, 'Come!' Let the one who is thirsty come; and let the one who wishes take the free gift of the water of life.'"

Comparing the beginning and the end of the Bible, it appears that the bride speaks with God at the end instead of only God Himself speaking at the beginning. What has happened? Who is this bride? How did God communicate to His bride in relation to His character and purpose of creation?

I am a physicist in training, and worked in the engineering field for thirty years. In analyzing a complex and challenging technical circuit or device, engineers usually use the *black box* technique. It is to find out the input and output first, while the in-between processes are temporarily represented by a *black box* as in Figure 11(a).

Figure 11a: Black Box Technique

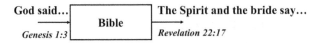

Figure 11b: Black Box Technique applied to understand the Bible's key messages.

This kind of analysis can help engineers to understand the main components that impact and operate within the *black box*. In the same token, the Bible appears to the general public as a complex book. A lot of people in developed countries have heard about the Bible's messages and stories. However, they are mostly in bits and pieces where the main theme of the Bible is often missed and misunderstood. There are also related legends and cultures that distort the original meaning of the Bible and make it difficult to comprehend.

For example, Christmas seems to relate to Christ, who is the original character of the Christmas stories. Now, *Santa Claus* replaces him as the main character. People in some Asian countries, think Christians believe in *Santa Claus* instead of Jesus Christ.

By focusing on the most important factors that acted as *input* for the Bible as in Figure11(b), you can know the real lifeline initiating the Bible. This lifeline is "God said." It connects all the people and events in the Bible. Therefore, "God" and "God said" represent the key essence to understanding the Bible. When "God speaks," He desires to communicate to someone who can receive His message and know what He's speaking about.

This path of communication from God continues all the way to the end of the Bible. As mentioned early in the *black box* analysis, the *output* of the Bible can be seen as what "the Spirit and the bride say" as in Figure11(b). You may ask: "Why is there a bride mentioned near the end of the Bible?" "What are the messages and meaning of 'the Spirit and the bride say?'" Once you have some understanding of these important characters and their representative comments throughout the Bible, key messages will be revealed.

You will find their exciting and glorious journeys to know God become more meaningful from the beginning to the end of the Bible.

The Scripture is full of examples about God desiring to have a family that expresses Himself. Jesus is not only the Son of God, but also the bridegroom of the bride at the end of the Bible. Such a beautiful picture is not a fairy tale like the prince and princess who live happily forever. The Bible is the actual communication to mankind about God's plan and processes to build His Holy family and Glorious kingdom from eternal past to eternal future.

As a scientist and engineer, I need evidence to back up whatever and whoever I believe in. Without the Bible, I would still be an atheist instead of a theist. Nevertheless, I have found my forty years and my intellectual effort to investigative the Bible trustworthy. Objectively, I can find strong evidence to verify the truth of what the Bible communicates to me. Subjectively, I can experience the spiritual relationship with Jesus Christ to validate His Word to me as true. I believe I am now a child of God and in His kingdom/family through Jesus Christ.

In these last two chapters, I want to walk through the wisdom within the communication of the special revelation of the Bible and Jesus Christ. They are full of divine wisdom and inspiration that impact mankind in different levels and areas. Jesus Christ actually is the living Bible who directly demonstrated and illustrated what it tells and teaches us. He is the center and main theme of the book.

So many reference books and interpretations of the Bible exist in literature, but the Bible is the top book for direct or indirect reference across different fields. Although the Bible, including the creation account in Genesis 1, is not a book to teach science, it is a book to teach wisdom[122]. When humans get this divine wisdom, they can then have the astuteness to develop science and art, and to build meaningful lives.

The impact of Biblical wisdom can be seen and discovered throughout history, even to this modern day. From art, architect, archaeology, science, society, and religions, human civilizations have developed around how people interpreted and understood the Bible. The first three items, art, architecture, archaeology, are more tangible items. They can touch the observer's heart, mind, and emotions, especially someone interested in learning more about the influence of the Bible. Art, architecture, and archaeology

---

[122] Dennis Prager, The rational Bible: Genesis, God, creation, and destruction, Regnery Faith, an imprint of Regnery Publishing, 2019: 8

are usually created at a specific time in history. Consequently, the people making historical art and architecture, and generating archaeological constructions—like the temple built by King Solomon—transformed their understanding of the Bible into expressive media at the time they lived. If you visit a museum that displays Middle Eastern and European art, you will see the visual impact of the artists' views of the Bible. Furthermore, the architect of church buildings in Europe, America, and even in some Asian countries, magnify Biblical treasures. The builders constructed the church buildings as sacred places for worshipping God, tending to express some form of artistic and technological thought based on their understanding of the Bible.

The background history of the Bible is mainly in the Middle East and its neighboring countries. They recorded the rise and fall of countries in the regions. For example, the Nineveh City of the Assyrian[123] Empire was buried in the sand after the decease of the empire. It was excavated by modern archaeologists to prove the objective truth of the Bible. What is important about archaeology to people reading the Bible is to show how it related Biblical faith to historical and cultural settings where Biblical events actually happened.

The faith of Christians around the globe is based on the actual person, Jesus Christ, who left His legacy and impact to the world around 2,000 years ago. Although He lived and ministered around a small area near Israel, Jesus Christ's influences have been spread across the five continents from the west to the east. Jesus' real-life events are recorded in the Gospels of the first four books of the New Testament, and also in other historical documents of the same period. One of the historical books was authored by the Jewish historian, Josephus. Flavius Josephus's Antiquities of the Jews, written around 93–94 AD, includes two references to the biblical Jesus Christ in Books 18 and 20[124] and a reference to John the Baptist in Book 18.[125] There were numerous findings in archaeology to verify evidence of the life of Jesus and His disciples in spreading the gospel throughout the regions of the Roman empire in the first and second centuries.

---

[123] Isaiah 10:5–19

[124] Josephus, Antiquities, 18, 20, Loeb Classical Library 456 (Cambridge, MA: Harvard University Press, 1965): 107–9

[125] Josephus, Antiquities, 18, Loeb Classical Library 433 (Cambridge, MA: Harvard University Press, 1965): 81–85. 9

The life of Jesus in His first coming to Earth started with a humble baby born in a manger. He grew up in a Jewish family and spent about three and a half years carrying out His Ministry to establish the kingdom of God. Jesus had the authority from God to teach, heal, and minister to people, demonstrating that He was the Messiah and Son of God. However, the Pharisees, Scribes, and Sadducees opposed Jesus's ministry because its teaching of Jesus threatened the groups' interests in maintaining their power, money, authority, and their wrong human expectation of the kingdom of God and Messiah.

Jesus confronted their idea that He is the True light to expose the darkness of humans' hearts, which had become spiritually blind. Eventually, the Pharisees and the Sadducees used political pressure to force the Roman official, Pontius Pilate, to execute Jesus in the cruelest historical way: by nailing him to the cross.

Jesus' sacrifice was to fulfill Isaiah's prophecy. "Yet it was the LORD's will to crush him and cause him to suffer... make his life an offering for sin... the will of the LORD will prosper in his hand."[126] His death on the cross was to pay the ultimate penalty of sin so He could save the people who believe in Him. The door to enter the kingdom of God was opened by His death and resurrection. The latter is the essential Biblical foundation of faith and hope for all Jesus's followers. It is this faith and hope that changes the hearts of mankind. In the spirit of what Jesus said in John 14:6, he is the only way to recover the lost communication between God and men. This is the real beauty of the Love of God, within the vivid record of the Bible.

Regarding my own experience reading the Bible, I received one as a gift from my friend about forty years ago in Hong Kong. I started reading from Genesis 1:1 but got stuck in understanding some of the Biblical stories related to Noah, the great flood, and the rainbow established by God as a sign, as recorded in Genesis 9:12–16:

> And God said, "This is the sign of the covenant I am making
> between me and you and every living creature with you, a
> covenant for all generations to come: I have set my rainbow
> in the clouds, and it will be the sign of the covenant between
> me and the earth. Whenever I bring clouds over the earth and
> the rainbow appears in the clouds, I will remember my cov-
> enant between me and you and all living creatures of every

---

[126] Isaiah 53:10

kind. Never again will the waters become a flood to destroy
all life. Whenever the rainbow appears in the clouds, I will
see it and remember the everlasting covenant between God
and all living creatures of every kind on the earth."

As soon as I read this story of the rainbow, I thought to myself the
rainbow belonged to nature and could be explained by the law of physics
in optics. I did not realize the relationship or the explanation as to why God,
as opposed to a natural phenomenon, could miraculously place a rainbow
in the sky. At that point, I stopped reading the Bible and returned it to my
friend. Thinking the Old Testament seemed to be nonscientific superstition,
I became biased and decided I did not want to be involved in any religion.
I did not do any further investigation of the evidence and facts related to
Biblical truth until I came to United States in 1981.

When I look back, I liked to share my mistake—my superficial understanding of the Bible and doubting God's mighty power. I now encourage
people to read and meditate on the Word of God with an open mind. As
mentioned by John Lennox, the Christian mathematician of professor at
Oxford University,

Regarding nature as only part of a greater reality, which
includes nature's intelligent Creator God, gives a rational
justification for belief in the orderliness of nature... in order
to account for the uniformity of nature, one admits the existence of a Creator, then that inevitably opens the door for
the possibility of a miracle in which that same Creator
intervenes in the course of nature... Surely, then, the open-
minded attitude demanded by reason is now to investigate
the evidence, to establish the facts, and to be prepared to
follow where that process leads, even if it entails alterations
to our *a priori* views. We shall never know whether there is
a mouse in the attic unless we actually go and look![127]

The turning point for me in appreciating the greatness of the Bible came
after I encountered Jesus through the four Gospels in the New Testament.
Opening my spiritual eyes, I understood there is no conflict between the

---

[127] John C. Lennox, God's undertaker: has science buried God? Oxford: Lion, 2009:
205–206

Bible and science. The Bible can complement the origins that science cannot answer, such as the Big Bang, the origin of biology and humanity. Although I still do not fully understand all biblical truth, it is enough for me to accept this truth with faith, based on the evidence I trust and my beliefs that the original authors of the Bible and the book of nature are the same Creator.

Since I accepted Jesus Christ as my Savior and established an eternal relationship with God, I love the Bible as a "living book." Through daily reading and mediating on the Word of God in the Bible, I can experience peace, joy, and hope even in this uncertain age's environment of complex political and social upheaval. All sixty-six books of the Bible are beautifully composed. Although the period written about in the Bible is about 2,000 to 3,000 years ago, the biblical wisdom and principles are still closely related to today's living. People from different groups, races, and countries still can apply Biblical wisdom and truth to build brighter and better lives. When human beings follow the manual of the book of life, they can find not only real peace and joy on a personal level, but also in their communities. Beside the impact on art, architect, archaeology, and science, the Bible also strongly influenced societies and religions:

First, let me talk about the most popular publication. The Bible is recognized as the book with most translated languages, and has the highest annual printing volume year after year, more than any other best-selling book in history. Due to e-books and Internet usage in recent years, millions of people have access to the Bible, and more peoples' lives have been affected. This is because the Bible feeds us spiritual food that can bring peace and joy through the Holy Spirit. Although the Bible has encountered degrees of prohibition of circulation in certain countries throughout history, people are even more eager to learn more about the Bible during the persecution of their faith.

In the last century, the Chinese government banned the circulation of the Bible. However, since the open-door policy started around 1980, Chinese people have gained understanding of the Bible and know the faith related to Biblical teachings is not an "imperial opium" for spiritual addiction. Instead, the Bible can inspire people to do good, with much "positive energy." The Bible is now generally available in mainland China, so there are no restrictions to reading it. It can also be printed and published there, making the Bible even more popular.

Second, let's look at the Gospel's positive impact on life. The gospel presented in the Bible is not a rigid, black and white doctrine, but a true light that shines on the human heart and spirit. The ambassadors of

Christ—evangelists—are not only verbally sharing salvation messages, but witnessing to others through their testimonies and good deeds. They are caring for orphans and widows, caring for patients in hospitals, bringing hope to prisoners, and comforting desperate and distressed people.

In addition, the genuine gospel transforms the lives of gamblers, drinkers, and people with drug addictions because Jesus's love and authority are able to rebuild people's physical, psychological, and spiritual lives. After their healings, people can return to society and make contributions to their communities. As the Apostle Paul shared in Romans 1:16, "For I am not ashamed of the Gospel, because it is the power of God that brings salvation to everyone who believes: first to the Jew, then to the Gentile."

A third idea to consider is the fulfillment of prophecy. The Bible is also a book of prophecies. The rise and fall of kingdoms—such as the prophecies of the Babylonian, Persian, Greek, and Roman Empires—have been recorded in Daniel 2, written about 2,700 years ago. God revealed to King Nebuchadnezzar through dreams and visions about what would happen to the kingdoms of the world at the time of Babylon and afterward. One vision showed him an enormous, splendid statue with a head made of pure gold, its chest and arms of silver, its belly and thighs of bronze, its legs of iron, and its feet partly of iron and partly of baked clay. Daniel interpreted the sequence of materials to represent the kingdoms of Babylon, specifically with King Nebuchadnezzar as the head of gold[128]. Babylon fell to the kingdom of the Medes and the Persians [129] afterward. Greece became the successor to the Medo-Persian Empire[130]. After Greece, Rome became the "iron" empire to rule during the time of Jesus Christ[131].

What he prophesied about the kingdoms was fulfilled one by one, which proves the truth of the Bible as the revelation of God. There is also a heavenly kingdom in the end of the pre-Christian era, with Lord Jesus Christ as the real King who will last forever: "In the time of those kings, the God of heaven will set up a kingdom that will never be destroyed, nor will it be left to another people. It will eventually replace all those earthly kingdoms with the heavenly kingdoms which will itself endure forever."[132]

---

[128] Daniel 2:38

[129] Daniel 5:26-31

[130] Daniel 2:39; 8:20–21

[131] Daniel 2:40–43

[132] Dan 2:44

Only the Creator of the heaven and Earth, who rules the history of the past and future, can precisely predict future events. "Beyond the evidence for the Bible's correctness (manuscript evidence) and its historicity (archeological evidence), the most important evidence is that of its inspiration. The real determination of the Bible's claim to absolute inspired truth is in its supernatural evidence, including prophecy. God used prophets to speak and write down His Word and He uses miracles like fulfilled prophecy to authenticate His messengers.

For example, in Ezekiel 26, we can see in astonishing detail how the city of Tyre was to be destroyed, how it would be torn down, and how its debris would be thrown into the sea. When Alexander the Great, king of Greece, marched on that area, he encountered a group of people holed up in a tower on an island off the coast near there. He could not cross the sea, so he could not fight those in the tower. Rather than wait them out, the proud conqueror had his army throw stones into the sea to build a land bridge to the tower. It worked. His army crossed the sea and overthrew the occupants of the stronghold. But where did he get so much stone? The rocks that were used for the land bridge were the leftover rubble from the city of Tyre… its stones cast into the sea!"[133]

Other similar prophecies can be found throughout the Bible and verified by checking historical facts and archeological evidence.

The greatest prophesies in the Bible are about the coming of Messiah or Christ from the first to the last book in the Bible. There are close to 300 prophesies related to the life of Christ. Most of them were already fulfilled through the first coming of Christ and His ministries near Israel about 2,000 years ago. The rest of the prophesies will certainly be fulfilled again in His second coming.

As Jesus Christ told His disciples about the sign of His second coming and of the end of the age. "Nation will rise against nation, and kingdom against kingdom. There will be famines and earthquakes in various places. All these are the beginning of birth pains."[134] No one on Earth will know exactly when Christ's second coming will be, but the signs of the end times are beginning to match what is happening in recent history. We know Christ's return grows closer and closer.

I am convinced that the occurrence and development of the prophecies in the Bible will be completed according to the date set by God. The

---

[133] https://www.gotquestions.org/which-book.html, accessed Aug. 13, 2019; Ezekiel 26:12

[134] Matt. 24:7–8

fulfillments of prophecies in the past and the foretelling of future Biblical events will bring real hope and meaning for Christ followers in building an eternal relationship with God. The kingdom of God is not only in the future, but also happening in the lives of believers today. Christ loves us first and was willing to sacrifice Himself in the cross to take away the blockage, sin, of our communication to God. Jesus Christ is now communicating with us in the spirit through the Bible. This living book helps us to walk with Christ according to the living principles of the kingdom of God in this world until the glorious return of the real King of Kings, Jesus Christ.

You can also find the most treasured gift within the limited period of human life, which has a birthday as the beginning and a death day as the end of human living on Earth. When the limited time you have is invested wisely, you and I can generate valuable rewards and spiritual fruit[135] in exalting and glorifying God through Jesus Christ. The fruit of the Spirit is "love, joy, peace, forbearance, kindness, goodness, faithfulness, gentleness and self-control"[136] that can build up people's character with integrity to do good works. This is the real beauty and purpose of human life that leads to faith, hope, and love in the present age to last into the eternal future, when God's love and truth will be fully revealed. The Bible, God's Word, is the most important key to opening the gate to His kingdom. Whereas Jesus Christ said, "I am the gate; whoever enters through me will be saved. They will come in and go out, and find pasture... I am the good shepherd. The good shepherd lays down his life for the sheep..."[137]

Are you ready to meet and know the good shepherd, Jesus Christ? Here are two books to further explore as informative tools in understanding the Bible.

First, *A Visual Theology Guide to the Bible: Seeing and Knowing God's Word*, combines graphics and text to teach the nature and contents of the Bible in a fresh and exciting way. The genuine truths of the Bible can be viewed and accessed in a way that can be seen, understood, and learned more easily than through traditional text-only methods. It is not only a brilliant introduction to Biblical values; it is a functioning guide for understanding and communicating divine wisdom and applying it to human living. You

---

[135] Galatians 5

[136] Galatians 5: 22–23

[137] John 10:9–12

can visualize and experience the wisdom of life in the Bible. As concluded by the authors, Tim Challies and Josh Byers:

> The Bible is completely trustworthy because it was breathed out by a trustworthy God who never lies. Jesus is the main character of the Bible's story, and the center of all that is taught and communicated. Creation displays the glory of Jesus. Mankind's fall displays our need for Jesus. The law sets the foundation for Jesus. The prophets proclaim the coming of Jesus, and the gospels reveal the coming of Jesus. The apostles' writings show a world being transformed by Jesus, as his Spirit continued his work and his church reflects his gospel. The end of the Bible shows that all things are culminating in the worship of his name.[138]

The second book is *Invitation to Biblical Interpretation: Exploring the Hermeneutical Triad of History, Literature, and Theology,* for those willing to study and research the Bible in a more advanced setting. It is one of the textbooks used in the class of Biblical Hermeneutics offered in Grand Canyon University (GCU). I attended the online class and completed the graduate certificate of Biblical foundation from GCU. The professor, Bob Hunter (Dr. B), together with resources from the university, provided students with effective learning guidelines and discussion forums. I gained more depth and insight about studying the Bible through Dr. B's stimulating questions in class discussion, his coaching for assigned homework, and by writing research papers for the class. The textbook uses an approach to the study of Scripture that properly balances history, literature, and theology.

The hermeneutical triad (Figure 12), which includes historical study, will prove to be a useful guide for mastering the general skills required for biblical interpretation and for following the special rules applied to the various genres of Scripture. Rather than being pitted against one another, history, literature, and theology each have a vital place in the study of the sacred Word. Regardless of the passages of Scripture, the interpreter needs to study the historical setting; the literary context including matters of canon, genre, and language; and the theological message — what the passages teach

---

[138] Tim Challies, Josh Byers, *A Visual Theology Guide to the Bible: Seeing and Knowing God's Word*, Zondervan March 26, 2019: 200

regarding God, Christ, salvation, and the need to respond in faith to the Bible's teaching.[139]

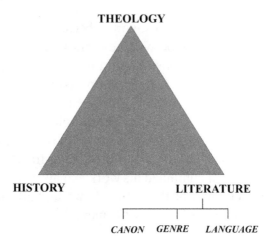

Figure 12: The Hermeneutical triad

---

[139] Andreas J. Köstenberger, Richard D. Patterson, *Invitation to Biblical Interpretation: Exploring the Hermeneutical Triad of History, Literature, and Theology*, Kregel Publications, Grand Rapids, MI., 2011: 78–79

Chapter Thirteen

JESUS CHRIST

J esus said, "The words (*Rhema*) I have spoken to you—they are full of
the Spirit and life[140]." His instant spoken words (*Rhema*) are related
to the living Word of God (*Logos*). These are two primary Greek words
in the Bible, which both translate to the same English "word" in the New
Testament. The first, *Logos*, refers principally to the total inspired Word[141]
of God, and to Jesus, who is the living Word in eternity. The second Greek
word translated "word" is *Rhema*, which refers to the spoken word. Rhema
literally means an utterance (individually, collectively or specifically)[142].
The original source of both *Logos* and *Rhema* are Jesus Christ.

Illustrating the principles for the eternal (constant) and instant-speaking
Word, I use an example of e-mail communication. Modern technologies provide and facilitate communication among two or more parties effectively
and almost instantly. The person transmits an email to another person as
receiver, who can read the message contents via e-mail on a computer or
smart phone as in Figure 13. However, the receiver needs the compatible
hardware and software to receive the email, assuming there are no blockages within the communication channel.

A recent instance of email communication may help to explain the
important of these communication cycles. I sent an e-mail (alantai@___.___)
with a sample writing document to my book editor, Ann Videan. She
received my email fine without problem and also replied back with an
email asking me if I received her email response. However, a blockage in
the email system did not allow me to receive her response. Unaware that
Ann sent me a response, I wondered if she was preoccupied and too busy to

---

[140] John 6:63b

[141] John 1–2; 1:14

[142] https://www.gotquestions.org/rhema-word.html, accessed Aug. 15, 2019

do some editing for me. In this situation, her email had already sent, and had existed for a constant period of time, but I did not receive it. She called me one morning to ask if I received her response. I told her I had not received it. She relayed the same message via our telephone call and asked me to use another email address for communication between her and me. Both sources—email and telephone—are from Ann. I followed her instructions and switched to the new domain address and was able to establish the correct communication both ways. I was able to receive her instant word to me when I spoke with her via telephone in real time, then via email later after I opened the email.

This example serves as an analogy about the communication between God and men through His Word.

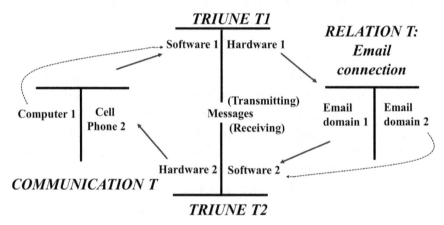

Figure 13: Example of computer and cell phone roadmap 2

Nevertheless, God is above in a domain of eternity (supernatural), not limited by time and space, but humans are limited here in their human domain (nature) with an end time on Earth. There is a limitation in using the example of email and telephone to illustrate divine wisdom. You need to apprehend and acquire wisdom yourself by reading the beautiful Word of God, the Bible. This establishes the communication channel with God through His instant Word when there are no blockages in your communication with Him. An impediment to Godly communication could be your strong opinion against Jesus, your biased thinking toward Christianity, your presupposed understanding of the Bible, or even human sin, as mentioned in the Bible. God already sent His beloved son, Jesus Christ (*Logos*), to the world; Jesus is the living word and living Bible to mankind and He has provided the only way for you to connect to Him, so you can listen to His instant spoken Word (*Rhema*).

His salvation through Jesus has existed for 2,000 years, since Jesus first came to this world. Now, it is your turn to connect with God by choice through your free will. Your prayer through Jesus is always received by God, but whether you can receive His spoken Word will depend on different conditions.

Using T transform in chapter six for some common communication between two parties, I applied similar approaches to create a roadmap for the communication between God and men. Going back to the creation described in Genesis 1–3, I started the roadmap in Figure 14 related to how God created man: Then God said, "Let us make mankind in our image, in our likeness, so that they may rule over the fish in the sea and the birds in the sky, over the livestock and all the wild animals, and over all the creatures that move along the ground." (Genesis 1:26)

Then the Lord God formed a man from the dust of the ground and breathed into his nostrils the breath of life, and the man became a living being. (Genesis 2:7)

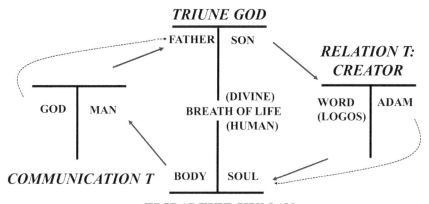

Figure 14: God and Man roadmap
Note: Then the Lord God formed a man from the dust of the ground and breathed into his nostrils the breath of life, and the man became a living being. (Genesis 2:7)

God's creation and making a man in God's image is by His Word. The breathing of God into the nostrils of man is the breath[143] of life, which is the key for a human being to be unique among all God's creations. No

---

[143] Breath in the original language, Hebrew, means spirit

other animal or plants require the breath of life from God or were made according to God's image and likeness as mentioned in Genesis. God talked to Adam directly, and Adam understood what God commanded him—not to eat from the tree of the knowledge of good and evil.[144] This is a representative picture of communication between God and men in the creation story recorded in Genesis.

As in Figure 14, the relationship between the triune God (Father, Son and Spirit) and man involved Him creating Adam by His Word. The breath of life from God connected Adam's Tripartite being—his body for the physical part, his soul for the psychological part, and his spirit for communication or fellowship with God. These three-in-one parts do not exist separately but are united as one human being. (See examples and the analogy of the computer back in Chapter 6 with its figures).

Since human beings were created in the image and likeness of God, we learned from the revelation of the Bible[145] that triune God is three-in-one—one God in nonseparating expression: Father, Son, Spirit. Through this special relationship between God and man, Adam was asked by God to rule over the fishes, birds, and animals with instructions and commandments about what Adam could or could not do in the Garden of Eden. Eve was also brought to Adam as helper by God through a special divine intervention.[146] This is the original kingdom of God where God created the heaven and Earth by His mighty power and divine wisdom. In His image man was made, God also gave man wisdom, free will, and authority to rule the Earth. There was communication between God and man, first through God's breath of life, then through God's Word to Adam for an intimate relationship.

The fall of mankind started when Adam and Eve disobeyed God's commandment. They listened to the false word from the serpent, which was actually the embodiment of Satan[147]. The serpent deceived Eve, as recorded

---

[144] Genesis 1:17

[145] The Father, Son and Spirit are mentioned directly in the New Testament of the Bible. (Matthew 28:19, Galatians 4:6, John 14:16). The triune God or Trinity is usually referred to as one God that eternally exists in three divine persons or expressions (equal in power and glory with the same attribute). The term Trinity is not directly mentioned in the Bible but is the theological terms that attempts to explain the mystery of God; https://www.blueletterbible.org/Comm/mcgee_j_vernon/eBooks/how-can-god-exist-in-three-persons.cfm

[146] Genesis 2:20–22

[147] Revelation 12:9; 20:2

in Genesis 3: "'You will not certainly die,' the serpent said to the woman. 'For God knows that when you eat from it your eyes will be opened, and you will be like God, knowing good and evil.' When the woman saw that the fruit of the tree was good for food and pleasing to the eye, and also desirable for gaining wisdom, she took some and ate it. She also gave some to her husband, who was with her, and he ate it."[148]

God told Adam that death was the consequence of eating from the tree of knowledge of good and evil. Though the physical bodies of Adam and Eve did not die right away, the Spirit part of fellowship and communication with God did not function after they disobeyed God's Word.

Professor Andy McIntosh further explained the fall's consequence, "As we read the terrible events of Genesis 3, we see what death is: separation. First there is spiritual death, as Adam and Eve know they are separated from fellowship with God, whereas before sin, they enjoyed perfect fellowship with Him... God then pronounces that Adam would die physically when He says, 'To dust you shall return.' (Genesis 3:19)"[149].

This result in death of life passed on to them and their offspring. After they were driven out from the Garden of Eden, the physical bodies of Adam and Eve died eventually. Altogether, Adam lived a total of 930 years, and then died, as recorded in Genesis 5:5.

---

[148] Genesis 3:4–6

[149] Andy McIntosh, What Is the Scriptural Understanding of Death? Creation and the Cross, May 23, 2016 https://answersingenesis.org/death-before-sin/scriptural-understanding-of-death/

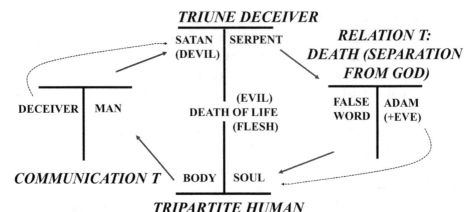

Figure 15: Deceiver and Man roadmap

Note: And the Lord God commanded the man, "…you must not eat from the tree of the knowledge of good and evil, for when you eat from it you will certainly die."(Genesis 2:16–17)

Figure 15 shows a picture summarizing the essence of the fall of mankind using the T transform. It started with the deceiver, the embodiment of Satan in the physical appearance of the serpent that distorted God's Word to Eve and Adam. The Hebrew word Satan means accuser or adversary and appears in the old Testament eighteen times[150]. It appears in the New Testament thirty-five times and means the devil, accuser,[151] and tempter[152]. The power of Satan lies in his tactic of leading the whole human race astray from God by his false word. God can certainly destroy Satan, who influences humans to do evil things.

Created in God's image and likeness, mankind has special privileges to choose to listen to the deceiver or not. The impact of the fall not only resulted in death, but also united mankind with Satan by listening to and believing the false word. The history of all mankind is full of terrible events done by murderers, traitors, thieves, liars, and numerous people doing evil things. Some of the killing, like that in the two world wars, did not happen in the animal kingdom among the same animal species. Men became the enemy of God by rebelling against God and disobeying God's Word.

---

[150] 1 time in 1 Chronicles 21:1, 14 times in Job 1 – 2, 3 times in Zechariah 3:1-2

[151] Revelation 12:9–12, "…the devil, or Satan, who leads the whole world astray. He was hurled to the earth, and his angels with him. …For the accuser of our brothers and sisters, who accuses them before our God day and night …"

[152] Mark 1:13a, "and he was in the wilderness forty days, being tempted by Satan."

God did not destroy and wipe out all human beings, because of His patience and love toward them. The dilemma of God is His righteousness in upholding the death penalty and His love in giving humans a chance to turn back to Him. In the same role as both a righteous judge and loving father, God would do whatever He could to save His unfaithful children. He developed a complete salvation blueprint to give hope to mankind.

God first spoke about a rescue plan or a gospel for future generations through the offspring of the woman (Eve) in Genesis 3:15. "And I will put enmity between you [serpent] and the woman, and between your offspring and hers; he will crush your head, and you will strike his heel."

During the Old Testament era and till the time of Jesus Christ in the New Testament, He continued to speak through the prophets to His chosen people, Israel, about the coming of the Messiah (Savior). Israel's ancestor, Abraham, was one of God's chosen people to receive the blessings, as mentioned in Genesis 18:17–19. Then the Lord said, "Shall I hide from Abraham what I am about to do? Abraham will surely become a great and powerful nation, and all nations on earth will be blessed through him. For I have chosen him, so that he will direct his children and his household after him to keep the way of the Lord by doing what is right and just, so that the Lord will bring about for Abraham what he has promised him."

Jesus Christ is the real blessing to all nations and is also the fulfillment of the promise to Abraham. He was not only the offspring of Abraham from a human genealogical point of view, but was also the Son of God, a direct expression of God in carrying out the salvation plan 2,000 years ago. Jesus Christ was also introduced in Hebrews 1:1–3 to both the Jews and all other people:

> In the past God spoke to our ancestors through the prophets at many times and in various ways, but in these last days he has spoken to us by his Son, whom he appointed heir of all things, and through whom also he made the universe. The Son is the radiance of God's glory and the exact representation of his being, sustaining all things by his powerful word. After he had provided purification for sins, he sat down at the right hand of the Majesty in heaven.

Jesus did not die directly by the hand of Satan, but by the hand of the rebellious people who did not apprehend God's Word and His salvation plan. It is the sin of all human race that Jesus bore His suffering on the cross. He

is God incarnate in human flesh, as the woman's offspring, to pay the ultimate death penalty for the sin of mankind inherited from Adam and Eve.
Although the sin corrupted mankind in following the devil and doing evil,
the only salvation to bring back life was Jesus Christ sacrificing Himself
on the cross. His shedding of blood has the divine power to cleanse sin so
we can reconcile with God. Hebrews 9:22 tells us: "In fact, the law requires
that nearly everything be cleansed with blood, and without the shedding
of blood there is no forgiveness." The law in the Old Testament required
the regular use of animals in sin offerings. However, we have been made
holy through the sacrifice of the body of Jesus Christ once for all,[153] per the
New Testament.

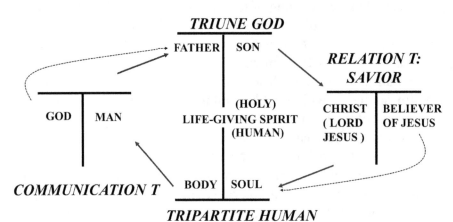

Figure 16: God and Man roadmap 2

Note: I Cor. 6:17, But whoever is united with the Lord is one with him in spirit. (or in
the Spirit)

The accomplishment of God's salvation plan through Jesus Christ, Son
of God, is summarized in Figure 16. It started with the triune God: Father
as the source[154], Son as the expression, and through the life-giving Spirit
(or Holy Spirit) to bring back life to those willing to receive and believe
in Jesus Christ. The tripartite human believer through the spirit, soul, and

---

[153] Hebrews 10:10

[154] 1 Corinthians 8:6, "yet for us there is but one God, the Father, from whom all things
came and for whom we live; and there is but one Lord, Jesus Christ, through whom all
things came and through whom we live."

body[155] can then reconnect and establish an eternal relationship with God. His or her body (ears or eyes) receives the gospel message into the soul (mind, emotion, will) so faith can be generated to receive the Savior, Jesus Christ, into his or her spirit. Jesus answered the question of a Pharisee named Nicodemus about being born again: "Very truly I tell you, no one can enter the kingdom of God unless they are born of water and the Spirit. Flesh gives birth to flesh, but the Spirit gives birth to spirit."[156] Therefore, human beings can be born again spiritually through Jesus Christ because of His death and resurrection. This is the only way in opening the door for believers to enter the kingdom of God and receive eternal life[157].

It is amazing human beings can have real hope and a bright future in God's kingdom, as mentioned in Colossians 1:12–14, "And giving joyful thanks to the Father, who has qualified you to share in the inheritance of his holy people in the kingdom of light. For he has rescued us from the dominion of darkness and brought us into the kingdom of the Son he loves, in whom we have redemption, the forgiveness of sins."

God's eternal salvation plan does not stop in the spiritual redemption (justification) of the believers, as in Figure 16, but has the complete salvation plan for the soul (sanctification) and body (glorification). The spiritual born-again believers need to grow in spiritual life just like the baby needs to grow up healthily through nutritious food and exercise. The spiritual food the believers need is the Word of God in reading the Bible and prayer in communicating with God. The spiritual exercise the believers need is the sharing of Christ to others and serving in His church. Without healthy spiritual food and exercise, there is no transformation of the soul in connecting to Christ or healthy growth.

Believers who genuinely receive Jesus as their savior are born again and become the children of God. This eternal salvation is only obtained as a gift by the faith of the believers, without works. However, spiritual maturity requires effort in coming to God with a believer's heart, soul, mind, and strength through the power of the Spirit, sent by Jesus Christ[158]. This could involve a daily internal struggle to spend quality time in reading and

---

[155] 1 Thessalonians 5:23b, May your whole spirit, soul and body be kept blameless at the coming of our Lord Jesus Christ.

[156] John 3:5–6

[157] John 3:16

[158] John 15:26

mediating on the Word of God. Externally, believers need to balance time so they can attend church meetings and serve in the church, even when they have a busy schedule. There are constant temptations for modern-day Christians to skip Sunday worship, fellowship, and prayer meetings because they are busy with work or entertainment, they deem more fun than church meetings. Therefore, the Bible has different verses to encourage the believers to put more effort in the transformation of their souls.

Two of these verses are:

"Do not conform to the pattern of this world, but be transformed by the renewing of your mind. Then you will be able to test and approve what God's will is—his good, pleasing and perfect will." (Romans 12:2)

"And we all, who with unveiled faces contemplate the Lord's glory, are being transformed into his image with ever-increasing glory, which comes from the Lord, who is the Spirit." (2 Corinthians 3:18)

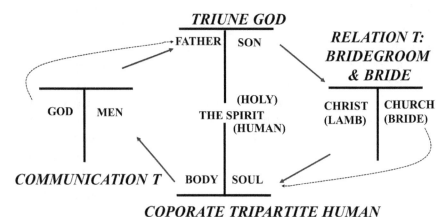

Figure 17: God and Man roadmap 3

Note: Reference to Revelation 19:7,"Let us rejoice and be glad and give him glory! For the wedding of the Lamb has come, and his bride has made herself ready."

The completion of God's salvation plan through Jesus Christ, the Son of God, is summarized in Figure 17. It is similar to Figure 16 as it starts with the triune God and the emphasis of building a more intimate relationship between Christ and the church. The word "church" is a translation of the Greek word *ekklesia*, which means "an assembly" or "called-out ones" of the believers. The Bible even uses the intimate relationship between husband and wife as an example to illustrate God's desire to build an eternal relationship with His church as a bride, through Christ as the bridegroom. An earthly marriage cannot be built into a loving relationship if the husband

and wife do not spend quality time together. Similarly, the heavenly marriage will be developed into a fruitful relationship when the church is built up and ready as in the heavenly visions given to Apostle John in the Book of Revelation: "I saw the Holy City, the new Jerusalem, coming down out of heaven from God, prepared as a bride beautifully dressed for her husband... 'Come, I will show you the bride, the wife of the Lamb.' And he carried me away in the Spirit to a mountain great and high, and showed me the Holy City, Jerusalem, coming down out of heaven from God."[159]

John's vision in Revelation includes the wedding feast[160] of the Lamb (Jesus Christ) and His bride (the church). Apostle Paul also refers to this spiritual union between Christ and the church as mystery being revealed. "For this reason a man will leave his father and mother and be united to his wife, and the two will become one flesh." This is a profound mystery—but I am talking about Christ and the church.[161] Note that this marriage relation of "one flesh" refers to the Genesis 2:21–24 that Eve and Adam shared the same composition physically and formed one complete union. Whereas, the common term of "flesh" refers to the human body that emphasis the human nature after the fall of mankind.

Figure 18 is an illustration of the union between a husband (Adam) and wife (Eve). As "one flesh," a married couple is united in their body (physical) and soul (mind, emotion and will) through love and respect. In the biblical view of "one flesh," it relates to the marriage of a husband and wife unified in sharing lives together not only in body physically but also emotionally, intellectually, and spiritually. The Bible has good teachings for maintaining a loving marriage, which is the most intimate relationship between two people. This relationship was used as metaphor for the spiritual union between God (Christ) and Men (church). Just like Adam and Eve are two individuals with a special relationship of husband and wife to become one flesh; God and Men form a special relationship of Christ and church to become one in the Spirit. As I Cor. 6:17 tells us that, "But whoever is united with the Lord is one with him in Spirit."

---

[159] Revelation 21:2, 9b–10; The scripture tells the readers about the work of the Spirit, while the Holy Spirit, the life-giving Spirit, the Spirit and the comforter are the same Spirit of the Trinity. The spirit of mankind can communicate with God through the Spirit during different experiences in his or her journey of spiritual growth.

[160] Revelation 19:7–10

[161] Ephesians 5:31–32

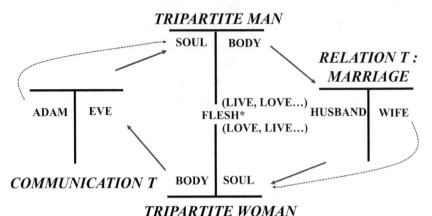

Figure 18: Adam and Eve roadmap

*Note: Matt 19: 5 ... the two will become one flesh)

This special relationship is for the church (corporate believers) to build up an intimate union with Christ for oneness of mind, soul, and spirit in the heavenly kingdom. The kingdom of heaven, which is also the kingdom of God, is for Christ, as King, to rule His heavenly kingdom so His children grow spiritually into maturity.

Matt. 22: 1–14 tells us that "Jesus spoke to them again in parables, saying: 'The kingdom of heaven is like a king who prepared a wedding banquet for his son....'" Mature believers will be invited to enjoy all the promise of blessings in a feast with Christ as bridegroom in His kingdom.

There will be a crown of life[162] and reward for those faithful believers doing God's will in building and preparing His church to be ready for the heavenly wedding. The book of James 1:12 says, "Blessed is the one who perseveres under trial because, having stood the test, that person will receive the crown of life that the Lord has promised to those who love him." The implication is that the believers need to love the Lord continually regardless of what environment they encounter, including life's trials.

However, the trial will end in Christ's second coming with the hope of glorification as revealed in Philippians 3:20–21, "But our citizenship is in heaven. And we eagerly await a Savior from there, the Lord Jesus Christ, who, by the power that enables him to bring everything under his control, will transform our lowly bodies so that they will be like his glorious body." This is the ultimate hope for all the genuine believers and the final consummations of the glorification in God's salvation plan. The plan is not a theory

---

[162] 2 Timothy 4:8, James 1:12

or on the drawing board for planning, but has been carried out throughout human history. Many believers willing to establish an eternal relationship with God through His Word have entered the kingdom of God and kingdom of Heaven[163], which showed the divine wisdom in saving people. The beauty of eternal life has been communicated to mankind through Jesus Christ.

Are you able to receive the messages and good news (gospel) while communicating back (prayer) to the wisest Creator and beloved Redeemer?

---

[163] There are two major theological interpretations related to the kingdom of God and the kingdom of heaven: 1) They refer to the same kingdom, ruled by God; 2) They refer to different times or spheres of the ruling by God. My own interpretation based on the mathematical set and subset definition is that the kingdom of heaven is a subset of the kingdom of God. All the spheres from figs. 14, 16 and 17 can be referred as the kingdom of God in general but specifically fig. 17 can be referred as the kingdom of heaven, which is a subset of the kingdom of God.

PART SIX

CONCLUSION

Chapter Fourteen

THE COMBINED ROADMAP

I conclude this book with a roadmap and the visions that touched me most in realizing the wisdom of science and life. Motivated by these visions as my missions, I want to share with you my experiences and understanding of the communication through the divine wisdom in this book. The combined roadmap in this conclusion integrates the maps of the Creation, Fall, and Salvation plan (Figures 14 to 17), which can be used, together with the Bible, for you to find treasure in life.

This roadmap provides you with a map to study the Bible from a communication perspective. It served as my Biblical guide to help me know God better once my spiritual eyes were opened forty years ago. I found the essence in the notes I wrote during that period made it easier for me to understand the Truth. One of the first Biblical notes I wrote in 1983 is shown in Figure 19. I saved this inspired prayer I'd written: "I know the truth is TRUTH. How wonderful for God's purpose. By the Lord's mercy, one day I will present the TRUTH to those people [who] open their heart[s] to receive Jesus Christ as their savior. Amen! Thank YOU LORD."[164]

---

[164] Alan Tai, Prayer note on June 18, 1983

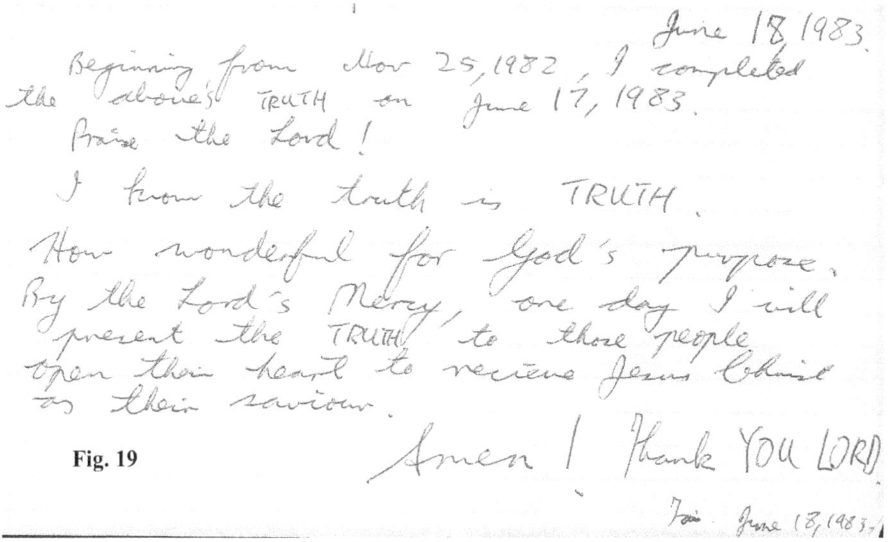

Beginning from Nov 25, 1982, I completed the "above" TRUTH on June 17, 1983. June 18, 1983.

Praise the Lord!

I know the truth is TRUTH.

How wonderful for God's purpose.

By the Lord's Mercy, one day I will present the TRUTH to those people, open their heart to recieve Jesus Christ as their saviour.

Amen! Thank YOU LORD

For June 18, 1983.

Fig. 19

Inspiration came to me from the Spirit during my study of the Bible. I started to record the first version of the roadmap (Figure 20) in November 25, 1982, at the back of a Bible. There were a few more versions after that as I spent more time studying the Bible to know Christ and to serve Him in church. It has been updated to the latest version in Figure 20.

As a physicist, I am more content to understand the fundamental relations among the key parameters of my scientific research and engineering applications. Studying the Bible required similar approaches to understand the Creator, who created the universe and established physics laws. In particular, Biblical terms with respect to their relationships with God the Creator, Savior, Church, kingdom of God, and so on can be connected for a completed vision of God's plan for mankind. What I wanted was to present all these tools in a way to help people find salvation in God through Christ the Lord. The maps are not important once you obtain the biggest treasure in Life—direct communication with the Creator! In reality, Christ is the power of God and the wisdom of God[165]. That is why building up a relationship with Jesus Christ and His church is a wise process. Each believer can participate in the kingdom of God.

---

[165] 1 Corinthians 1:22–24

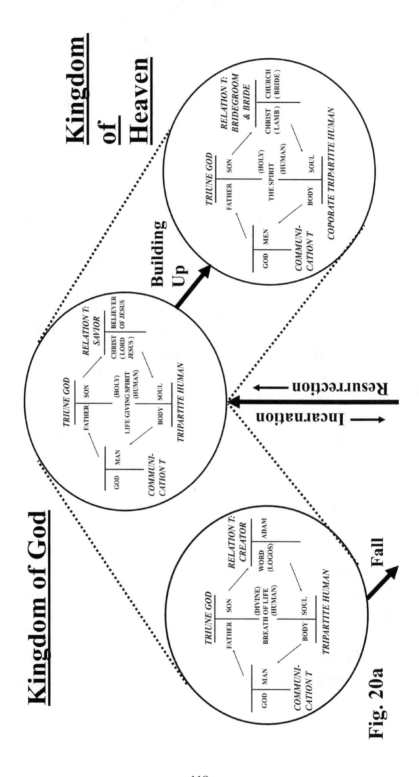

Fig. 20a

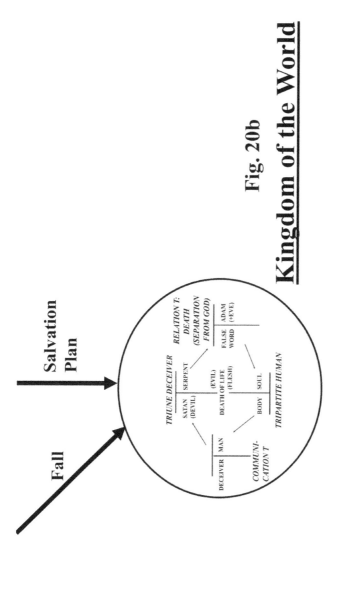

Salvation Plan

Fall

TRIUNE DECEIVER *RELATION T:*

DEATH (SEPARATION FROM GOD)

SATAN (DEVIL)   SERPENT   FALSE WORD   ADAM (+EVE)

(EVIL)

DEATH OF LIFE (FLESH)

DECEIVER   MAN   BODY   SOUL

COMMUNI- CATION T   *TRIPARTITE HUMAN*

Fig. 20b

Kingdom of the World

Figure 20 (a + b): God and Man combined roadmaps

The left circle (Figure 14) in Figure 20a represents the relationship between the Creator (triune God) and Adam, who became a living being[166] through the breath of life from God. However, the Fall brought in death (separation from God) when Adam and Eve acted according to the deceiver's false word, as in Figure 20b (Figure 15). From life to death, the tragic results not only impact Adam and Eve, but also pass on death to all human beings. Satan brought evil and darkness to the kingdom of the world so the history of mankind became full of fighting, killing, lying, and sinning. There was no way out of the bondage of sins even after the law of God was given to Moses, or the moral teachings by religions. People rebelled against God repeatedly. Is there hope for human beings? Why are there so many human disasters and so much sufferings? Where is the righteousness and savior of the world?

God had a salvation plan to save human beings through Jesus Christ. Conceived by the Holy Spirit and born from a woman, Mary, Jesus' first coming to the world was through incarnation: making Him fully divine and fully human. The arrow connecting the center circle in Figure 20a to Figure 20b represents His incarnation to live in the world and then die on the cross to accomplish God's salvation plan. From humanity's point of view, Jesus was the first human resurrected from the dead. From the divine point of view, God's mighty power raised Jesus from the dead to justify human after the penalty of sin was paid by Jesus' death. His redemption opened the gate for human beings to enter the kingdom of God. This is represented by the arrow connecting the circle (Figure 15) in Figure 20b to the center circle (Figure 16) in Figure 20a. Human beings can be born again spiritually by receiving Jesus Christ as their Savior, so the life-giving Spirit can be united with the believer's spirit through faith in Christ. As illustrated in John 20:22 after Jesus' resurrection, he breathed on the disciples and said, "Receive the Holy Spirit."

Romans 5:17–19 gave a precise summary to the righteous act of Jesus' death:

> For if, by the trespass of the one man, death reigned through
> that one man, how much more will those who receive God's
> abundant provision of grace and of the gift of righteousness
> reign in life through the one man, Jesus Christ! Consequently,
> just as one trespass resulted in condemnation for all people,

---

[166] Genesis 2:7

so also one righteous act resulted in justification and life for all people. For just as through the disobedience of the one man the many were made sinners, so also through the obedience of the one man the many will be made righteous.

Regarding Jesus's resurrection in Romans 6:59:

For if we have been united together in the likeness of His death, certainly we also shall be in the likeness of His resurrection, knowing this, that our old man was crucified with Him, that the body of sin might be done away with, that we should no longer be slaves of sin. For he who has died has been freed from sin. Now if we died with Christ, we believe that we shall also live with Him, knowing that Christ, having been raised from the dead, dies no more.

The right circle (Figure 17) in Figure 20a represents the building up of His church through the Spirit, sanctifying the believers and renewing their minds after they are born again in Christ. The building up of God's spiritual dwelling place requires the church's (corporate believers') enduring efforts to prepare themselves till Christ's second coming. This building up is represented by the arrow joining the center circle to the right circle, kingdom of Heaven, as in Figure 20a. There are other Bible verses that help us to understand more about the amazing salvation plan carried out through Jesus Christ. The readers are encouraged to study the Bible directly to gain more insight into God's heart's desire for you to build an eternal relationship with Him.

This roadmap is just one of the tools you can use to establish your own theology in knowing God. At the end of the day, the Bible tells us, "The kingdom of the world has become the kingdom of our Lord and of his Messiah, and he will reign for ever and ever."[167] I believe this will happen as revealed in the Bible, and I presented it in the roadmap tool to help as a bird's eye view for anyone interested in studying the Word of God.

It may take some time for you to embrace the complete picture of these roadmaps. I trust it will help you understand the relationship we discussed in the last chapter and in this conclusion chapter. Your comments and notes can be added to the roadmaps while reading the four Gospels of Jesus Christ

---

[167] Revelation 11:15b

or other books in the Bible. Writing down your own comprehension of the Bible, gained through prayer and the guidance of the Holy Spirit, is always the right thing to do. You can even modify or improve the roadmaps to gear it up with your own perspective in understanding God's salvation plan based on Biblical Truth.

Most of the followers of Christ agree upon a few absolute Truths about Christ, along with the traditional view of Biblical Truth. Here are a few of them for your reference and verification, with respect to the roadmaps and the Bible. Christ followers believe Jesus Christ is the Son of God, fully human and fully divine, and that they receive eternal life by believing in Him and following His teachings. Jesus Christ died on the cross for humanity, God raised Jesus from the dead, and He will come again at the end of time. Also, the genuine believers believe, "there is one body and one Spirit, one Lord, one faith, one baptism, and one God and Father of all, who is above all, and through all, and in you all"[168].

---

[168] Ephesians 4:4 -5; the emphasis of oneness in these Bible verse.

Chapter Fifteen

TRIUNE GOD AND YOU

The triune God or Trinity is a mystery, but the revelation of the Bible tells us "God the Father from whom all things came and for whom we live, God the Son or Redeemer, Jesus Christ, through whom all things came and through whom we live"[169], and God the Holy Spirit is to comfort, console, and guide those who belong to Christ[170].

I cannot fully understand the mystery of the Trinity, but the roadmaps help me to realize the relationship somewhat clearer after reading and mediating on the Bible for years. The more I spend time in God's Word, the more I come to know Him, not just in knowledge but also in realty of experience with the triune God.

Without the living Word of God, He could be very abstract, and people could misunderstand Him. Using a real-life example of understanding how the radio works, I will demonstrate the general principle of turning misconception into the truth.

When I first heard and saw a radio at a young age, I was curious about it. Someone told me little people inside the radio made the sound. I was really not sure if that was the truth until I opened the cover and found only components inside. No little people. However, I still did not know what caused the radio to make the sound.

As I grew up, I read books about the radio and also attended the Technical Institute in Hong Kong, studying electronic engineering—including radio and television—for two years full-time and four years part-time. I learned the theory behind the radio and even built a real one to gain a better understanding about it. However, I still did not understand what moved the communication through the media between the transmitter and receiver. I went to schools in the United States to learn the physics of radio waves (electromagnetic waves) and spent ten years studying, from undergraduate through

---

[169] 1 Corinthians 8:6

[170] Isaiah 11:2; John 14:16; 15:26; 16:7

graduate school. It gave me a better picture and understanding about how the radio works, but still saw more mysteries in the electromagnetic waves. However, the more I studied the radio and related subjects, the more I appreciated it. I also was able to apply the understanding to my work as an engineer and scientist.

Similarly, with the study of the Bible, we need to open it and systematically study and research it to know the living Word. It is important to find out yourself if what you think and perceive about God is in agreement with what the Bible reveals to you. The more you spend time on the Word of God, the more you will find more new insight and concepts you did not think about before. Certainly, God wants you to know and apply His Word to experience Him in real life so you can build an eternal relationship with Him, the Creator and the Redeemer. Once you have the divine communication channel and relationship established, your application of His Word will bear spiritual fruit and you will witness God's glory in your life.

People use different talents when learning about science and art. This makes them see and experience life from different angles. As in Figure 21, mathematics, physics, chemistry, and science belong to the objective and measurable group while art, humanity, sociology, and language belong to subjective and artistic group. I used to favor my interest in the scientific side while lacking interest in the artistic side. Focusing on mathematics and science in my studies and career, I lacked the humanity and language skills to communicate with others. On the other hand, some people have talent on the artist side while lacking interest in the scientific side. There could be conflict and difficulty in understanding one another given the perspectives between two distinct types of people.

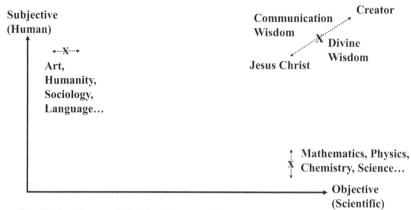

Figure 21: Objective and Subjective graph

The Creator of the heaven and Earth can bring harmony to all His creation as demonstrated in the beauty of sky, ocean, trees, flowers, and so on. Different sciences lie behind the planets and biological life, which provide opportunities for human beings to understand and study the wonders. Meanwhile, the many artistic visages in all creatures allow human beings to appreciate and share the wonders. The Creator possesses divine wisdom, like a perfect scientist and a perfect artist, harmoniously balancing both science and art. Jesus Christ is the expression of God and Prince of Peace[171] to bring harmony among different types of people in this world.

In addition, God is a speaking God. When He speaks, creation and order follow. If God did not speak at the beginning, nothing material would exist. There will be no heaven and Earth. More importantly, there would be no human beings, and you would not have existed. Have you wondered why you exist? It is because God continued to speak right after the "beginning." Therefore, order and light replace void and darkness when God speaks. Because God speaks, the order, science, and life spring up and shine forth as the beautiful creation in the world.

In God's creation, the human being is the most "beautiful" creature He ever created as we were created in His image[172]. God want us to listen to His speaking so we can communicate with God to build an intimate relation with Him eternally. However, the disobedience of Adam and Eve

---

[171] Isaiah 9:6

[172] Genesis 1: 26–27

blocked the communication channel between the entire human race and God. There was only one way to resume His communication with human beings. By His wisdom of speaking through Jesus Christ, any human being willing to receive Christ can have new life in the new creation. This new creation replaces the old creation, which was under the condemnation of God because of Adam's and Eve's disobedience. However, "there is now no condemnation for those who are in Christ Jesus"[173]. This is the key for all the believers to come back to the heavenly Father through the Son of God, Jesus Christ, by the power of the Holy Spirit. All those people in Christ Jesus can be united back to the triune God through their faith in Christ.

Since I opened my heart to Jesus Christ, the Creator and Redeemer, I realized He is not only full of divine power and wisdom but also full of humanity in loving and caring for people around Him. He was also a great communicator in relating to people the right ways, truth, and life pertaining to the kingdom of God. The world holds billions of people who believe in Jesus and follow Christ. His life also transforms me to be more balanced toward the understanding of science, art, humanity, and life. This is one of the motivations for me to write this book, to share what I learned over the years of walking with Christ and studying His Word. I hope this book will give you more appreciation of the divine wisdom in science and life.

Billy Graham (1918–2018), studied the Word of God for more than eighty years and brought the gospel around the world. Rev. Graham proclaimed that, in the wonders of nature, we see God's laws at work. Who has not felt astonished and humbled at the glory of God's handiwork when looking up at the night sky? Who has not felt uplifted seeing all creation burst with new life and vigor when spring comes each year? While we see the magnitude of God's power and His infinitely detailed planning in the beauty and abundance around us, nature does not directly tell us of God's love or grace. Our conscience only tells us of the presence of God and the difference between good and evil. It is in the lessons within the pages of the Bible that we find the clear and unmistakable message which true Christ followers are based upon.[174]

The creation and Biblical messages resonate with my experience and also with many others' experiences. I applaud the wisdom in my studying of science and understanding of life while appreciating the divine wisdom in the Bible. God has been communicating with me since I accepted Jesus

---

[173] Roman 8:1a

[174] https://billygraham.org/devotion/wonders-of-nature, accessed Aug. 15, 2019

as my Savior. He is the Creator and the Lord of this universe. I hope you can also realize His communication as beautiful and wonderful so you also experience the joy and peace of knowing Jesus Christ.

The prayer below is just an example of how you can communicate with God if you are not so sure how to talk to Him. Feel free to use your own words to represent what you want to ask and talk from your heart.

"To the Creator of this universe,

I would like to know you even though I am not so sure how. I see the wonder of the universe and the goodness of nature provided to me. Your Love and Grace through Jesus Christ really touches me.

Please help me to break the communication barrier, my sin, through the salvation of Jesus in the cross. Your Spirit helps me to connect to you as I open my heart to receive new life, in Jesus' name."

From the sincere prayer of _____ (your name)

Angels[175] and myself (Alan Tai)
rejoice with you for your new life in Christ.

May God bless you in your new journey in the kingdom of God.

    Growing your spiritual life in Christ:
1) Communication:
    a. You communicate with God through prayer.
    b. God communicates with you through your reading of the Bible.
    c. You communicate with brothers and sisters in Christ through sharing in small groups and worshipping in church.
    d. You communicate with non-Christians by witnessing Christ through words and works with wisdom. (Col. 4:2–6)
2) Love:
    a. Love God[176].
    b. Love yourself and your family.
    c. Love your brothers and sisters in Christ.
    d. Love one another. (I John 4: 7–21)

---

[175] Luke 15:10

[176] Matthew 22:37

I dedicate the following poem (Figure 22 in Chinese) to those who love the Truth, and "ask God to fill you with the knowledge of his will through all the wisdom and understanding that the Spirit gives, so that you may live a life worthy of the Lord and please him in every way: bearing fruit in every good work, growing in the knowledge of God," (Colossians 1: 9b – 10)

See the majesty of the heaven and earth,
that my heart has long adored;
Examine the wisdom of the nature's law,
that my mind is deeply fascinated with.

Vastness of universe is all ruled by the Lord of lords;
For the living God reveals the Truth to me.
May people turn their hearts to Jesus;
God's beloved Son shed blood to cleanse their sins.

Peace and joy will fill their lives abundantly;
God's amazing grace will bless them endlessly;
Sing praise and glory to the living God!

觀宇宙穹蒼奧秘，察自然定律神奇；

尋主宰掌管萬物，求活神啟示真理；

願人回轉信耶穌，神子流血洗罪過；

得平安喜樂一生，享永福神恩無窮。

榮耀頌讚永生神！

戴志忠
2019年8月7日
鳳凰城

Figure 22: Poem for science and life

128

BIBLIOGRAPHY

Bradstreet, David and Steve Rabey. Star struck: seeing the creator in the wonders of our cosmos, Grand Rapids, Michigan: Zondervan, 2016.

Challies, Tim and Josh Byers. A Visual Theology Guide to the Bible: Seeing and Knowing God's Word, Zondervan, 2019.

Collins, Francis. The Language of God: A Scientist Presents Evidence for Belief, Simon & Schuster, 2006.

Dao, Chu (Y C Ruan). Life Made in Wisdom: the Mathematical Principles of Biointelligemce and the Origin of Life, Xulon Press, 2018.

Gates, Bill. The Road Ahcad, Penguin: London, Revised 1996.

Giorgio, Philip. Creation and the Arrow of Time, Xulon Press, 2017.

Guillen, Michael. Amazing Truths: How Science and the Bible Agree, Zondervan, 2016.

Hutchinson, Ian. Can a scientist believe in miracles? InterVarsity Press, 2018.

Köstenberger, Andreas J. and Richard D. Patterson. Invitation to Biblical Interpretation: Exploring the Hermeneutical Triad of History, Literature, and Theology, Kregel Publications, Grand Rapids, MI., 2011.

Lennox, John C. Gunning for God: why the new atheists are missing the target, Oxford: Lion, 2011.

Lennox, John. God's undertaker: has science buried God? Oxford: Lion, 2009.

Lennox, John. Seven days that divide the world: the beginning according to Genesis and science, Zondervan, 2011.

Nathan Aviezer. In the beginning: Biblical creation and science, Hoboken, N.J.: Ktav Pub. House, 1990.

Nee, Watchman. The Normal Christian Faith, Church in Hong Kong Bookroom Ltd. Company, 1984.

Parker, Andrew. The Genesis Enigma, London: Doubleday, 2009.

Polkinghorne, J. C. The God of hope and the end of the world, New Haven: Yale University Press, 2002.

Prager, Dennis. The rational Bible: Genesis, God, creation, and destruction,

Regnery Faith, an imprint of Regnery Publishing, 2019.

Ross, Hugh. Beyond the Cosmos, RTB Press, 3rd edition, 2017.

Ross, Hugh. Improbable planet: how earth became humanity's home, Grand Rapids, Michigan: Baker Books, 2016.

Ross, Hugh. The Creator and the Cosmos: How the Latest Scientific Discoveries Reveal God, RTB Press, 4th edition, 2018.

Shalev, Baruch. 100 Years of Nobel Prizes, Americas Group, 2005.

Spitzer, Robert J. New proofs for the existence of God: contributions of contemporary physics and philosophy, Grand Rapids, Mich.: William B. Eerdmans Pub., 2010.

Strauss, Michael G. The Creator Revealed: A Physicist Examines the Big Bang and the Bible, Westbow Press, 2019.

Tai, Alan Chi-Chung. Study of unconfined states in quasi-periodic semi-conductor superlattices, Ph.D. Thesis, Boston College, 1991, University Microfilms International, Order Number or ProQuest publication number: 9211799 (https://dissexpress.proquest.com)

Tai, Alan. http://ccmusa.org, search 戴志忠: 10/01/2015 兩本最有智慧的書, 戴志忠 (中信月刊), 05/01/2017 一本生命的活書, 戴志忠 (中信月刊)

Appendix I

GENERAL COMMUNICATION VIA HTTP://WWW.SCIENCEANDLIFE.NET

Through the advance of Internet technologies for communication, we can read and connect with others across the globe without the boundaries of countries and communities. There is some background information related to this book on the Internet at my personal site www.scienceandlife. net. Please feel free to browse the pages and posts for background information that sets the stage for my writings. It will be updated from time to time, and I will respond to some of the meaningful comments as needed.

I am a physicist in training and I am writing this book related to the wisdom of science and life. It is a response to my passion about learning science and life throughout various stages of growing into an adult. I present here, a bit of the background from the virtual book that serves as introductory information for this book.

The information presented there should help you understand the processes of discovery of science, a world view, and value of life. Certainly, all humans want to live meaningful lives influenced by their beliefs, including scientists and religious leaders.

Even though some of them no longer exist on this Earth physically, their contributions and spirit for seeking the truth continues to impact the world. If you would like to share insights about their successes and lives, please feel free to post your impression and knowledge of them in the comments area of www.scienceandlife.net. Your feedback to me about this book is also encouraged via alantai@scienceandlife.net.

Please use the Internet and email to share productive ideas, opinions, and books etc.; and I ask you not to include commercial products and political promotions.

I'm striving for the continuation of meaningful dialogue. As long as we have a heart to respect others, the communication can create an atmosphere in which to appreciate and realize the wisdom in science and life.

Regardless of your background, you will encounter all or some of the questions in Table 1 while you live on this Earth.

Table 1 Questions encountered and possible answers in your life

Questions encountered in your life	Example of a general answer	Another possible answer	Opposite of the general answer example
Does God exist?	Yes	Maybe	No
Who is God?	Jehovah	Do not know	Different answers
What evidence supports the above questions?	Bible	Do not care	Not need
How to communicate with God?	Prayer	Not sure	Nothing
Where is God?	Everywhere	Somewhere	Nowhere
When I can communicate with God?	Anytime	Sometime	No time
Why do I live in this time and space?	A mission from God	Somehow	By chance

There could be broad spectrum of answers to the questions. Trying to list the possible range of the answers in the table, I respect your answer because of your own reasoning or background.

On the other hand, I believe people should respect one another regardless of their background, races, beliefs, and ages. If we can understand more about other people's situations and promote dialogue, I believe the world will have more peace and fairness.

There are so many scientists and different beliefs I could choose for the dialogue on my web site. However, Marie Curie, Albert Einstein, and Isaac Newton are the scientists who had the character and spirit of investigation which really resonated with my inner being. They were my role models for great scientists when I was still a student in Hong Kong.

Another group of people with different beliefs are also included. Having a heart of respect for others with different traditional religions, I attended various religious schools in elementary and high school. The atmosphere and teachers in the schools impacted my education and childhood/life in Hong Kong.

I grew up in a grassroots community where my parents held a belief of traditional Buddhism. I even went to school with a Buddhist background for first grade. Failing in most subjects except mathematics, I only studied in the school for one year. I was then enrolled in second grade at a Catholic school and finished my elementary education there. I failed the Chinese Language subject in my secondary school entrance examination, which made it difficult to enter high school. However, I passed the entrance test to the Islamic high school near my apartment and was allowed into their program. After I finished high school, I became much more interested in science than anything to do with religion. I kept searching for the meaning of my life while living in Hong Kong. Eventually I got a chance to study abroad and my horizons were opened… to discover more about science and life.

Appendix II

INTRODUCTION OF T TRANSFORM FOR MATHEMATICAL FRACTION

his appendix shows only the concept and high-level introduction of T transform used in mathematics as mentioned in chapter six. No complicated mathematical equations are involved here while more detail explanations will be presented elsewhere. Example of decimal number expressed to fraction, which is very special case for T transform, is used as demonstration for the application of T transform. Its equivalent value does not mean other terms (parameters) in other applications to be always equal, so precautious need to be used in relating similar expressions to different applications.

There are different ways to express 0.407407407407407... beside the method of decimal point number. Fraction can be one of the ways to express it with the special case of equal to 11/27 in mathematical terms. Figure 23 shows the decimal point number can be transformed through a special relation of reciprocal and rearrangement of the decimal point and digit. Note that this special case of expressive transformation is related to equal but most other cases need not be equal but just expression of the original term.

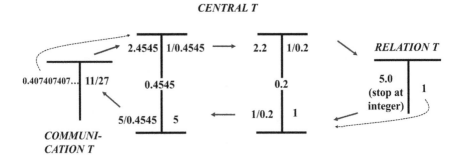

Figure 23: Example of mathematical T transform roadmap

The central T in Figure 23 stopped after 2 terms of transformation in this example. It can be continued for more terms depend on the starting value of decimal point number. A generalized equation can be combined all the terms in the central T to only one term (like the T transform in Figure 3) that will be easier for computation and analytical presentation.

Continue fraction is another way to convert the decimal number to fraction such as below equation of 3 terms of computation.

$$0.407407407\ldots = 1/(2+1/(2+1/5)) = 11/27$$

In general, the example presented here can be computed also using continued fraction for different rational numbers. I used computer program to verify the T transform technique can be used for the approximation of irrational number like pi, $2^{1/2}$ etc. In these cases, the expressions are not equal but approximation or close to the original term.

For less accurate approximation with only 4 terms of the computation using T transform,

$$Pi \sim 3.141592 = 355/113 = 3+1/(7+1/(15+1/(1)))))$$

For more accurate approximation with 13 terms of the computation using T transform,

$$Pi \sim 3.14159265358979 = 80143857 / 25510582 =$$

$$3+1/(7+1/(15+1/(1+1/(292+1/(1+1/(1+1/(1+1/(2+1/(1+1/(3+1/(1+1/(14)))))))))))))$$

ACKNOWLEDGMENTS

First and foremost, I want to thank God for His providence and love throughout my journey to wisdom in life. Knowing God and the laws of physics behind His creation, I find it the greatest treasure to work to unlock the mysteries of this universe.

I am grateful to my late advisor, Professor Goldsmith for his guidance and coaching for my Ph.D. thesis in quantum physics. As a great educator, he had the wisdom to motivate students to be curious and persistent in doing research. I am also thankful to my spiritual family at the Greater Phoenix Chinese Christian Church, that gives me the opportunity to preach and teach Biblical Truth. The pastoral staff, elders, brothers, and sisters in Christ have been a strong spiritual support to me in writing this book through their prayers and encouragement. Special thanks to Revs. K.F. Yang, Peter Liu, Semson Nip, and Elder Albert Wong for the fellowship and practicing the spiritual truth.

My special thanks go to Dr. B. (Bob Hunters) for teaching Biblical foundation classes and writing the endorsement for this book. His deep understanding of the Bible and stimulating questions helped me to write better research papers to meet the certificate requirements at Grand Canyon University.

I am very much thankful for my beloved wife, Priscilla, and my dear daughters, Breanna, Victoria, for their understanding, prayers, and continuing support of me while completing this book.

I express my thanks to Rev. Stephen Leung and Steve Miller for their insightful book endorsements related to the Christian faith and science.

I thank the following individuals for their assistance and professional tasks throughout different aspects of preparations for my manuscript and book. Ann Videan, my book shepherd, helped me in editing and providing insight so that I could write with clarity toward the goal of my book. Erica Coulter, pre-production rep., and Marc Bermudez, project coordinator, supported me in coordinating different phases of publishing my book through Xulon Press. Greg Dixon, editorial consultant, Jason Fletcher, book sales rep. and Jason Shingleton, marketing sales rep., book cover design team and the typesetting team gave their professional services from Xulon Press.

Thank God for all of you who impacted me in my growth, learning, working and writing of this book. My parents, relatives, teachers, coworkers and friends have all contributed to different degrees of shaping me in pursuing my life mission to know God and let God be known. Just to name a few of you: Joe Yeung, Ted Clifford, Bob Robert, Steve & Lily Kwiatkowski, Gerald & Beryl Chan, Rev. Paul & Ruth Li, Rev. Wilford & Elizabeth Kong, Gilbert & Christina Yip, Kent Tong, James Shieh, Antonio Chan and so on.

ABOUT THE AUTHOR

A lan Tai received his PhD in physics, specializing in quantum well research, from Boston College, and a graduate certificate of completion in biblical foundations from Grand Canyon University. Alan has worked at various medical ultrasound technology companies, including Philips as an engineer and GE Healthcare as a scientist. He has managed new medical ultrasound transducer products from development conception to successful commercial release. During his career in the engineering field, Alan has eleven issued patents and two pending patents. He is passionate in sharing how science and nature are in harmony with the creation processes revealed in the Bible. Being a community scholar in RTB (Reasons to believe, *www.reasons.org*), Alan joins the scholars to present the good news from the Scripture with sound reason and scientific research, which provides evidence in the truth of the Bible and faith.

CPSIA information can be obtained
at www.ICGtesting.com
Printed in the USA
BVHW040811100121
597337BV00002B/3